HUMAN BIOLOGY
AND ECOLOGY

WITH CONTRIBUTIONS BY

WINTHROP A. BURR, M.D. Clinical assistant professor of psychiatry, Tufts University School of Medicine

JONATHAN S. FRIEDLAENDER, Ph.D. Associate professor of anthropology, Temple University

DONALD E. GERSON, M.D.

WILLIAM WHITE HOWELLS, Ph.D. Professor of anthropology, Harvard University

ROGER REVELLE, Ph.D. Richard Saltonstall Professor of Population Studies, Harvard University, and professor of science and public policy at University of California, San Diego.

HUMAN BIOLOGY AND ECOLOGY

by ALBERT DAMON

Foreword by W. W. Howells

ILLUSTRATED BY ROMAYNE TIMMS

W · W · NORTON AND COMPANY · INC · NEW YORK

ACKNOWLEDGMENTS:

Page 28 (Figure 2.7): "Adaptations to Cold" by Laurence Irving. Copyright © 1966 by Scientific American, Inc. All rights reserved.

Page 29 (Figure 2.8): *The Evolution of Man* by H. W. Magoun. The University of Chicago Press. Copyright © 1960.

Page 33 (Figure 2.13): *Vertebrate Paleontology* by D. Romer. The University of Chicago Press. Copyright © 1967.

Page 36 (Figure 2.14): "The Antiquity of Human Walking" by John Napier. Copyright © 1967 by Scientific American, Inc. All rights reserved.

Page 66 (Figure 4.12), Page 164 (Figure 11.1), Page 183 (Figure 12.2): *Heredity, Evolution, and Society, 1st Edition,* by I. Michael Lerner. W. H. Freeman and Company. Copyright © 1968.

Page 79 (Figure 5.7): "Sexual Demorphism in Interphase Nuclei" by M. L. Barr. *American Journal of Human Genetics.* The University of Chicago Press. Copyright © 1960.

Page 131 (Figure 9.2): Using data on adopted children from Skodak, M. and Skeels, J., *Journal of Genetic Psychology*, **75,** 1949.

Page 133 (Figure 9.3a,b,c): *Principles of Human Genetics, 3rd Edition,* by Curt Stern. W. H. Freeman and Company. Copyright © 1973.

Page 151 (Table 10.4): Original source, *A Textbook of Surgery, 6th Edition,* by John Homans, 1945. Courtesy of Charles C. Thomas, Publisher, Springfield, Illinois.

Page 154 (Figure 10.5): "The Human Population" by Ronald Freedman and Bernard Berelson. Copyright © 1974 by Scientific American, Inc. All rights reserved. "The Populations of the Developed Countries" by Charles F. Westoff. Copyright © 1974 by Scientific American, Inc. All rights reserved.

Page 195 (Table 12.5): *Mendelian Inheritance in Man* by V. A. McKusick. The Johns Hopkins University Press. Copyright © 1966.

Page 305 (Figure 19.1): "The History of the Human Population" by Ansley J. Coale. Copyright © 1974 by Scientific American, Inc. All rights reserved.

Library of Congress Cataloging in Publication Data

Damon, Albert, 1918–1973.
Human biology and ecology.

Includes bibliographies and index.
1. Human biology.   2. Human ecology.   I. Title.
QP34.5.D34   573   77-559
ISBN 0-393-09103-1

CONTENTS

FOREWORD
By W. W. Howells

Albert Damon developed this book from the lectures in his popular General Education course at Harvard entitled "Human Biology and Ecology." As he said, no such book existed. It expresses the viewpoints he formed from his unusual life and education, as well as his feeling of the need to teach the biology of man to undergraduates and others not necessarily interested in biology itself.

Only he could have written this particular book. His own undergraduate education was in sociology, but he took his Ph.D. degree in biological anthropology and followed it with an M.D. in general medicine. He remained a practicing and consulting physician all his life, but his research was in such anthropological fields as medical and constitutional anthropology and human engineering, ranging from how to design a safer seat for truck drivers to health and disease in Solomon Islanders. It was his blended training and experience that provided the essentially humanist view from which he wrote the book.

His broad professional contacts and his understanding of many fields led him to see some missing connections in several levels of education and career development. As to ordinary education, he stressed a rounded, integrated view of the biological side of human life, different from the lumps of hygiene, of frog-dissecting, of museum trips, and of just plain gossip which make up the information of many people. In medical training he noted the orientation toward producing people more and more skilled in treating the sick and dealing with specific diseases. But, he held, an increase in the number of people so trained was not the only promising way to reduce the economic and social burden of human illness. Answers

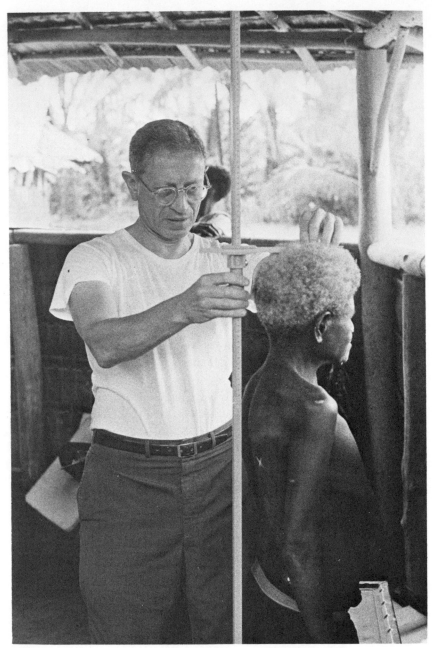

Dr. Albert Damon, Harvard University, Solomon Islands Expedition, 1970.

to fundamental problems, he said, would depend on fuller understanding of all the conditions under which people are likely to be long-lived, free of disability, productive, and creative. This should call for more young people to look into a wider range of careers—most of them in fact available but not generally known to college students—lying beyond medicine, medical research, and nursing (and all the necessary supports like administration and public health). These further careers would include biostatistics, epidemiology (the background of disease transmission), population surveys with relation to medicine, health economics, and so on. He wished to present human biology to college students as a way of drawing attention to these practical aspects, as well as to teach a general and integrated understanding of the biology of man as part of anyone's education for citizenship.

Before he died, Damon asked me to see the book into print, with the help of Peter Timms, who for several years had been his assistant in the General Education course from which the book derives. Without Damon's presence, I have felt particularly obliged to be chary of making changes in his style or statements, though I might disagree with occasional views or forms of expression; also, the book is in his competence, not mine. I have changed only what appeared (perhaps in the light of later information) to be actual inaccuracies, and have done some writing for the sake of smoothing and unifying the continuity. I have done some routine editorial work, of course. Otherwise, the text is Damon's, except for four chapters contributed by colleagues who usually covered the same topics in the course and to whom thanks are due. These colleagues include Jonathan S. Friedlaender, Winthrop A. Burr, Donald E. Gerson, and Roger Revelle. Dr. Timms and his wife Romayne selected points for illustration; she devised and executed the drawings and he wrote the legends. He also selected photographs for the plates, did the Glossary, and compiled the Suggested Readings for each chapter; and he substituted occasional updated tabular material for that originally used. Professor Edward E. Hunt, Jr., an old friend of both Damon and myself, reviewed the manuscript independently on behalf of the publisher and contributed comments, all of which were adopted. Others who reviewed the manuscript and contributed helpful suggestions were Vera King Farris; Walter C. Quevado, Jr.; Richard V. Bovbjerg; David G. Post; I. Lester Firschein; J. S. Weiner; Rada Dyson-Hudson; and Francis E. Johnston.

A couple of special editorial notes. The text lacks references to sources. This is because Damon died before being able to put all these in place. Many were present in the manuscript, and could also have been given for tables which do not show them. However, neither Dr. Timms nor I felt able to supply these consistently throughout, nor even to track down with certainty those which were present, so I have deleted them all, to arrive at consistency (and a more readable text) in this way.

A reader has drawn attention to the use of *man* and *mankind*, for *human beings* and *humanity*, a use which appears to reflect an age-old neglect of womankind that is unacceptable today and that many would like to exorcise linguistically. Such feelings are understandable, although the traditional usages are very hard to circumvent fully without being ungainly. Actually the ambiguity is ancient, from at least the time of Homer. *Anthropos* and *homo* are the Greek and Latin words from which the names of our science and our genus derive; and both have always definitely meant both "a man" and "a human being," the latter regardless of sex. Perhaps English seems more explicit, although the case is the same (German is ambiguous: *der Mensch* means "a human being," "a person," before it means "a man," while humanity, *die Menschheit,* is feminine in gender!) *Man* today can only mean the kind of animal we are, regardless of sex, just as when we talk about the domestication of horses, cows, ducks, and geese we certainly imply mares, bulls, drakes, and ganders.

At any rate, since I feel in good conscience that, when used in a general sense, *man = person*, I have not tried to expunge, on Damon's behalf, the words *man* and *mankind* when so used. (For myself, redemption would be rather too late anyhow.) I must say on his behalf that no subconscious acknowledgment of male priority was involved. Damon was too clear-headed, and was also an egalitarian in matters of race and sex, as he was in all matters except recognition of personal worth and allegiance to the search for truth.

HUMAN BIOLOGY
AND ECOLOGY

1
THE STUDY OF
HUMAN BIOLOGY

Human biology is a new science. Some scientists, in fact, boggle at the idea of a *human* biology because we do not deal in dog biology or fly biology. Quite right; except for a few matters of structure and chemistry, all the basics are *biology*. Any new discovery having some breadth of meaning reaches to man in the same instant as to other forms of life. DNA and the double helix, which burst on the scene less than a generation ago, are as much the monitors of developing human embryos as of tadpoles—there are no secrets that are human secrets only.

But none of this means that *Homo* is just another animal. We are the only creatures who are fully conscious of ourselves, the only creatures who even know there is such a thing as biology. We do not have to think of ourselves as the center of life, or lose sight of our continuity with other animals, to realize that the biological study of man is something different. Our needs and expectations, our whole view of ourselves, puts things on another plane. Human biology is the attempt to be systematic and scientific about the things that have always been the great concerns of man: life and death; health and sickness; being born, growing up, and getting old; and all the natural differences between every person and every other person because of genetic destiny.

Man is unusual or unique in various ways. He is bipedal and both sexes are almost hairless; these are significant, but they hardly justify a *human biology*—there are plenty of peculiar animals. We are the latest major kind of mammal to have evolved, which is more important. We have close relatives, but our life has rapidly become very different from theirs because of our behavior. We make tools and communicate by spoken language, and we recognize and formalize our social patterns, so that we can attempt conscious control of our contacts with nature and other human groups. This situation is the matrix of the biological concerns mentioned just above, and of the continuing problems we face. Man is a relatively sensitive animal, with a rich capacity to react and adjust, individually or in whole social groups, or, over several million years, in the evolution of the species itself. So, for human biology, the matrix is a skein of conditions involving evolution and variation, the appearance of human culture and technical achievement, and the effects of these on the ecology of man, past and present.

Evolution and Variation

Evolution. This book is not really about human evolution. But man's present nature cannot be understood without knowing what he owes to the past: his mammalian heritage and his kinship, through long-ago common ancestors, with those other brainy and sensitive animals, the higher primates—monkeys and apes. So, although the story often seems to be merely an obligatory chapter early in any book on man, it has basic significance. More important for us is evolution in general. Today it is the central, unifying principle of all the biological sciences. It has coordinated interpretations from every field of biology, for example paleontology, cell biology, and systematics. That is, things as disparate as fossil dinosaurs, DNA and the submicroscopic units of heredity, and the study of the dynamics and relationships of animal species, are all now brought onto common ground in evolutionary theory.

This body of theory has become very large and complex since Darwin's day (Figure 1.1). His contribution, which established evolution in science, came in providing the force necessary to make it work. This was natural selection working on heritable variation so that individuals with better natural, inherited variations of a feature would tend to be selected for survival and reproduction. Thus a population or a species would more and more take on the complexion of the most fit individuals. Much of Darwin's *Origin of Species*, in which his own contribution was developed, is in fact devoted to the matter of natural variation in animals. The important points are that, for evolution to take place, variation must be inherited in order to affect succeeding generations, and also that such variation must exist in the first place. No variation, no evolution. I shall have

1.1 *Vanity Fair* cartoon of Charles Darwin.

much to say about all kinds of variation in man in this book. But we should bear in mind that if man arose by evolution and natural selection, those features which are important in his present makeup must have shown natural heritable variation in the past for selection to work on.

Genetics and the Preservation of Variation. There was one thing Darwin did not understand: the mechanical basis of natural variation and the differences among individuals. In fact, the lack gave him much trouble. Why was not the progeny of two differing parents intermediate between the two in all ways, and so why did the variation not simply disappear? The answer was found by his contemporary Gregor Mendel in 1865, but Mendel's discovery lay dormant until 1900 when his work was recognized for what it meant. Only then began the science of genetics, which has become central in all biology and is, for example, the backbone of modern cancer research, as any newspaper constantly reveals.

The essence of Mendel's finding was what we now call genes; that is, the units of heredity which remain the same from parent to offspring.

They are unchangeable (normally) and incompressible. They do not blend. The differences come not from changed genes but from the *combinations of genes*. It is the combination, whether of the two genes in a single pair, or of all the genes in a person's makeup, that makes a new individual, an *individual* in the fullest sense of the word. The possible combinations of genes from two parents, let alone those in a whole population, are so nearly infinite in number that no person has ever had anything like a genetic replica—with the dramatic exception of identical twins.

The meaning of this for Darwinism and natural selection became clear. Because genes do not change (except for rare mutations) the variation in a population does not collapse. It is simply stirred. But natural selection can work by selecting favorable genes or combinations of genes, and so making them commoner in the population from generation to generation (for as long as such combinations continue to be favorable). Thus the biological nature of the population shifts, whether in visible things, like color, or invisible, like disease resistance.

For human biology there is a special significance in genetic heredity, which is probably still not well understood popularly. As in Darwin's day, the feeling seems to be that strength of heredity means being like one's parents. Of course one tends to be more like one's parents than like random members of the population, because all one's genes come from them alone. But that is not what a biologist means when he says that such and such a trait is under strong genetic control. He is saying that *differences* among individuals are due primarily to gene differences, not to accidents of development or environmental effects. We are perhaps too familiar with likenesses of brothers to each other or to their father—something which the averaging of many genes will indeed tend to produce. We should certainly bear more in mind the astonishing *differences* we usually see between brothers, in appearance or in mathematical or musical ability, and so on. That is the real meaning of genetic variation. It is in fact what Darwin would have much liked to know.

Human Variation. This is a principal field of anthropology (Figure 1.2) which deals with whole individuals in all their characteristics. It goes right down to molecular differences in proteins, but it is different from the study of anatomy or physiology; deviation from the textbook muscle, nerve, or reaction is simply a distraction to a teacher of anatomy or physiology. Anthropologists are interested in the whole array of differences of individuals and, equally importantly, of populations of individuals, across the probably four billion people now alive (see Figures 1.3 to 1.14 and 14.1 to 14.12). Anthropology, of course, is also interested in the evolutionary side: problems of long-term adaptation of populations—how such

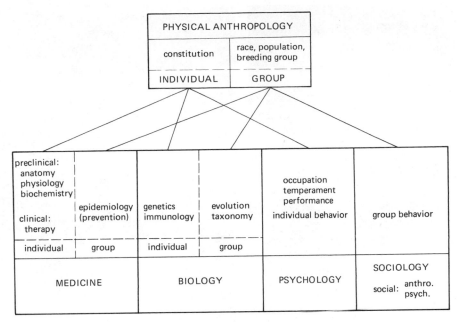

PHYSICAL ANTHROPOLOGY		
constitution	race, population, breeding group	
INDIVIDUAL	GROUP	

preclinical: anatomy physiology biochemistry clinical: therapy	epidemiology (prevention)	genetics immunology	evolution taxonomy	occupation temperament performance individual behavior	group behavior
individual	group	individual	group		
MEDICINE		BIOLOGY		PSYCHOLOGY	SOCIOLOGY social: anthro. psych.

1.2 Sources and relationships of physical anthropology.

groups, including major races, came to differ as they do by processes of evolution in recent millennia. We must bear in mind that evolution is not something we are finished with, because processes of population change are still going on. Such ongoing evolution can be illustrated, we shall see, by long-term trends in body size, length of life, age at menarche and menopause—that is, the onset of cessation of female reproductive life respectively—and the altered bases of natural selection in modern societies.

Human Culture
and Behavior

All animals "behave," and we should not forget that evolution and variation operate on behavior, not simply on bones or eyes. Horses, we know, evolved powerful legs adapted for running and high-crowned teeth adapted for feeding on the tough, abrasive grass of the plains. But it is how a horse takes advantage of these in actually running and feeding that determines evolutionary success. The primates, above all man, have developed alert senses and a relatively large brain; this intelligence of man has been a prime factor in his behavior and thus his progress. If a man can

1.3–1.10   Mongoloids: Orientals, Eskimos, and Indians.

1.3   (Left) Tibetan man. (courtesy Peabody Museum, Harvard University)
1.4   (Right) Mongolian man. (courtesy Peabody Museum, Harvard University)

1.5   (Left) Siberian woman. (courtesy Peabody Museum, Harvard University)
1.6   (Right) Eskimo woman. (courtesy Peabody Museum, Harvard University)

1.7 (Left) Hopi Indian girl. (courtesy Peabody Museum, Harvard University)
1.8 (Right) Dakota Indian woman. (courtesy Peabody Museum, Harvard University)

1.9 (Left) Arawak Indian man, northern South America. (courtesy Peabody Museum, Harvard University)
1.10 (Right) Araucanian Indian women, Chile. (courtesy Peabody Museum, Harvard University)

1.11 (Left) Man from South Australia. (courtesy Peabody Museum, Harvard University)

1.12 (Right) Maori man with face carving. (courtesy Peabody Museum, Harvard University)

1.13 (Left) Melanesian: Solomon Islands man. (courtesy Peabody Museum, Harvard University)

1.14 (Right) Polynesian: Samoan woman. (courtesy Peabody Museum, Harvard University)

ride a bicycle when he needs speed, he does not have to spend millions of years patiently evolving faster running legs but can make do with what he has. Early in his career as a human being, stone knives, hand-held clubs, and fire made him a more powerful hunter and fighter than his limbs or fingernails would have allowed by themselves. Much later inventions and skills permitted man, a tropical animal, to live very successfully in the Arctic, as do the Eskimos.

The Meaning of Culture. All these invented, outside appurtenances, from a club to the most intricate modern machine, are part of what constitutes human culture. Culture also covers the rest of man's patterned behavior, such as social institutions and religion. Other animals may form mated pairs and social groups, but man makes use of understood conventions to control these things, and he gives them names. Fundamental to cultural behavior, at least as we know it, is the essential cultural element of language. Language makes possible the storing of the cumulative experience of the species, so that useful discoveries of inventions can be kept by all succeeding generations. Once some genius discovered how to make a fire or use a wheel, mankind did not have to rediscover fire-making or reinvent the wheel every generation.

Culture is the web of learned custom which has become necessary to the organization of human economy and society. Culture mediates man's perception of the environment and his response to it. It may determine whether you regard a stranger as someone to eat or as someone to welcome as a ritual cousin. It determines whether you perceive a pig as embodying your ancestor's spirit, as an everyday source of food, or as an abomination you would rather die of starvation than eat. In 1972 a biomedical team from Harvard studied two tribes on isolated islands of the Solomon Islands archipelago in the Southwest Pacific. The waters surrounding each island teem with fish. One group, excellent fishermen, ate plenty of protein and were superbly nourished. The other, culturally indifferent to fishing and unskilled at it, ate mostly carbohydrates; the people were malnourished, with greatly depressed hemoglobin levels. Thus culture makes it possible to starve in the midst of plenty and, as we are all painfully aware, one need not go to remote parts of the world to make the discovery. Culture, in fact, determines who eats and who starves, who marries whom, how many children they have, who is exposed to what diseases, who survives, and who succumbs. Culture affects biology—but, in turn, biology affects culture. Let me give examples of each.

Culture and Biology. A good example of culture affecting biology is the distribution of the sickle-cell gene, hemoglobin S (Figure 1.15). This gene is found in up to 40 percent of the inhabitants of some tribes in the

1.15 The distribution of *HbS*, the gene which in heterozygote form produces the sickle-cell trait (but not the hemolytic anemia produced by the homozygote) correlated with the distribution of malaria. In malaria-prone lowland regions of Africa, the frequency of the *HbS* gene reportedly can reach 40 percent.

malarial regions of East Africa, its area of greatest concentration. In double dose, the gene is fatal; the homozygote, with two genes for *HbS*, dies at an early age, without reproducing. How can so lethal a gene persist in such high frequency? The answer is that the single dose confers some protection against the malaria parasite, which does not flourish in red blood cells with *HbS* (Figures 1.16 and 1.17). Pygmies and other African forest-dwellers have very low frequencies of *HbS* because the mosquito vector of malaria does not thrive in dense jungle. Only when immigrant farmers from West Africa many centuries ago began to clear the equatorial forests for agriculture were open spaces provided where stagnant water could accumulate, and where the malaria-carrying mosquito could breed. When the Africans were brought to North America, where malarial areas are minor, the frequency of individuals bearing hemoglobin S fell to about 9 or 10 percent and is continuing to fall.

    To illustrate the effect of biology on culture (Figure 1.18), consider the speeding up of human maturation that has been occurring over the past 150 to 200 years. People are not only growing larger, but are also reaching their final size and physical maturity much earlier. For example, girls in most of the industrial countries of the world now attain menarche (beginning of menstruation) between their twelfth and thirteenth birthdays on the average. This is some three years earlier than it was a century ago. In general, earlier physical maturation brings on a whole host of educational, social, and medical consequences, ranging from early marriage, reproduction, and divorce to school dropouts, venereal disease, and infant

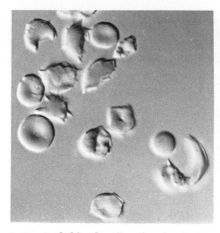 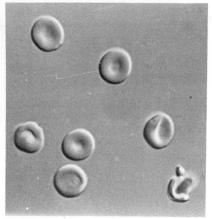

1.16 Red blood cells of individual with sickle-cell anemia. (courtesy Anthony Cerami, Rockefeller University. From *Proceedings of the National Academy of Science* 68: 1181. 1971.)

1.17 Normal red blood cells. (courtesy Anthony Cerami, Rockefeller University. From *Proceedings of the National Academy of Science* 68: 1181. 1971.)

and maternal mortality. The well-known generation gap, including school and campus unrest, may have as one of its roots the conflict between young people who today are biologically mature, and social rules devised for the immature adolescents of yesteryear.

There is other historical evidence of the interaction of biology and culture. During the Roman empire and at the time of the Renaissance, the age of menarche occurred earlier than it did during the nineteenth century. We lack precise records, of course, but Roman and Hebrew tomb inscriptions and literary accounts indicate that girls matured and married in their early to middle teens. In Shakespeare's *Romeo and Juliet*, Capulet says of Juliet,

> My child is yet a stranger to the world,
> She hath not seen the change of fourteen years.
> Let two more summers wither in their pride
> Ere we may think her ripe to be a bride (Act 1, Scene 2).

To this, her suitor Paris replies, "Younger than she are happy mothers made." And then Shakespeare shows a remarkable knowledge of human biology in Capulet's rejoinder: "And too soon marr'd are those so early made."

Infants born to teen-age mothers are indeed more likely to be prema-

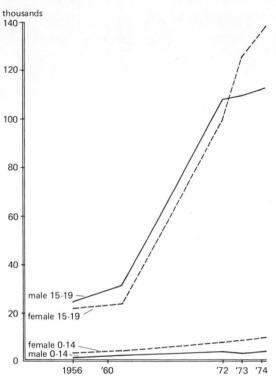

1.18 Biology affects culture. While the age of menarche has fallen, the reported cases of gonorrhea in the United States have risen steadily due to earlier sexual activity. (SOURCE: Department of HEW Center for Disease Control, Venereal Disease Control Division)

ture, undersized, and at greater risk of death than those born to women in their twenties and thirties. Later on, the Industrial Revolution is thought to have impaired living conditions to such an extent as to delay the average age of menarche, and possibly to depress height as well.

Human Ecology. This is the study of relationships among culture, habitat, and biology. Here the unit of study is the human population—its size, structure, dynamics, and relation to its environment with respect to food, energy, and material resources. Ecologic maladaptations take the form of disease, physical or mental; failure to reach genetic potentials, as in stunted physical and mental development; and various behavioral and social pathologies. Ecologic maladjustments take the form of population increase, environmental pollution, and overutilization of resources, coexisting with widespread starvation, poverty, violence, and disease.

There is extensive and justified concern for man's ability to survive in the human ecosystem. As we are well aware, this system contains, in addition to physical, chemical, and biological components, other men and their products. These products are cultural as well as biological, chemical, and physical. We are rapidly altering physical characteristics of our environment, to the detriment of the other organisms in it, and ultimately to our own hazard. Yet it is also evident that our present level of civilization and well-being has been attained largely through the manipulation of our environment and that any change in the way we handle our physical surroundings may have profound implications for the people who live in the real world. We all recognize the dangers of DDT, but if its use were stopped tomorrow, tens of millions of people around the world would die of malaria. This is no idle doomsday speculation; such a calamatiy actually happened in Ceylon in 1969. The extraction, transportation, and burning of fossil fuels does violence to the natural environment and imposes hazards to human health. Stop it for a few days, however, and civilization comes to a halt. Again, this is no doomsday prediction; it happened during the coal miners' strike in Britain during the winter of 1972.

A General View

There is much more to human biology than could be put in any single book, and I should like to say what I want to do in this one. I have barely suggested above some of the hazards that we and our descendants face. We know, from reading the newspapers, the kinds of solutions being sought. Some are heroic, some merely technical; some political, some scientific. Some give rise to fierce contention, like birth control; others, like organ transplants or genetic counseling, cause less controversy and seem merely like happy results of progress, when used according to directions. Many, like genetic engineering, or euthanasia versus prolongation of life by all possible measures, stir ethical debate. But they are not likely to affect the future of the species in any substantial way. In fact, they are nonissues, when one considers what a small part of the population they really affect.

I shall not try to provide the answers, big or little. I shall stick to the main concerns: the numbers and diversity of mankind, the biological effects of various environments on man, and the effects of man on his environment, which in turn affect his own biology and behavior. I want to provide an outline of fundamental knowledge of human biology, with a theoretical framework that will allow a reader to make sense of facts as he or she meets them, and with an approach that will help to evaluate problems and proposed solutions.

SUGGESTED READINGS

Coon, Carlton S., ed. 1948. *A reader in general anthropology.* New York: Holt.

Damon, Albert. 1975. *Physiological anthropology.* New York: Oxford University Press.

Dobzhansky, E. 1964. *Mankind evolving.* 5th ed. New Haven: Yale University Press.

Harris, Marvin. 1975. *Culture, people and nature.* New York: Crowell.

Harrison, G. A., Weiner, J. S., Tanner, J. M., and Barnicot, N. A. 1964. *Human biology.* New York: Oxford University Press.

Mering, Otto von, and Kaskan, Leonard, eds. 1970. *Anthropology and the behavioral and health sciences.* Pittsburgh: University of Pittsburgh Press.

2
MAN'S PLACE IN NATURE

The earth and solar system were formed about 4.6 billion years ago, by the condensation of interstellar matter. When the earth had cooled enough, an atmosphere formed. Heavy elements, like iron and nickel, settled into the center of the mass, or the core, and the lighter elements became concentrated near the surface. These lighter substances included hydrogen and oxygen, which form water; carbon, the distinctive element in the organic molecules which constitute living tissues; and nitrogen, which is the distinctive element in protein. Protein is the basic structural component of the body and forms many of its fundamental components as well. Evidence from the study of volcanoes and the relative frequency of elements suggests that the early atmosphere of the earth was made up of water vapor, methane or marsh gas, ammonia, and free hydrogen gas. This sort of atmosphere is still found on the planets Jupiter and Saturn. Over the ages, hydrogen has been used up and oxygen produced so that the atmosphere of the earth that we breathe is roughly 80 percent nitrogen, 20 percent oxygen, and virtually no free hydrogen.

Giant Early Steps

Laboratory experiments have shown that energy from various sources, abundantly available at the beginning, can transform simple molecules into larger ones. The four main sources of energy are: heat, produced by the condensation of matter; occasional lightning, from electrical

discharges in the atmosphere; and constant radiation—ultraviolet
radiation from the sun and cosmic radiation from outer space. The simple atoms and molecules in the primordial broth, when subjected to
these sources of energy in the laboratory, and presumably at the
beginning of life on earth, form more complex molecules—namely proteins, fats, and carbohydrates, which are the building blocks and fuel of
the cells and the body. The same chemical reactions produced the nucleic
acids, which transmit the hereditary message from one generation to the
next.

   The Cell. The most important step toward life was the enclosure of
complex molecules within a membrane (Figure 2.1). A surrounding membrane protects its contents from destruction. It keeps them in close contact with each other and thus increases the chance for chemical reactions
to take place among them. It can transmit some kinds of molecules while
acting as a barrier to others. It has been possible to study, in the laboratory, stages through which large molecules form aggregates, droplets, and
primitive membranes; but the key event, formation of membranes, is not
nearly so well understood as the steps from elements to molecules and
from molecules to aggregates.

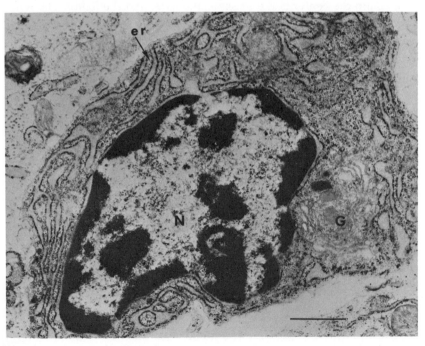

2.1   A plasma cell enlarged 20,000 times as seen through an electron microscope.
(courtesy WHO, Monkmeyer Press)

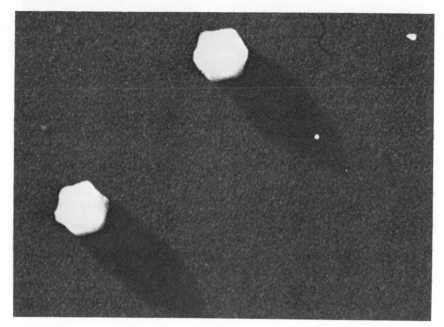

2.2 Electron micrograph of *Tipula* iridescent virus. (courtesy Virus Laboratory, University of California, Berkeley)

Nucleic Acids. In all living organisms the genetic information is carried by nucleic acids, in the form of DNA or RNA. (These things will be fully explained later.) They are called nucleic acids because they occur in the cell nucleus, which contains the chromosomes and genes; and, in fact, the genes themselves are composed of nucleic acids.

They act as a template to reproduce molecules of the specific kinds needed, and aggregates of molecules. The nucleic acids became the major organizers of molecules very early, probably about four billion years ago, in a step which was the beginning of life. The closest things to naked genes found in nature today are the viruses (Figure 2.2), which are nucleic acids enclosed in a thin shell of protein. Presumably they represent an extremely early stage in the development of life; fossil algae and bacteria, single-celled organisms which are a higher stage, have been identified in rocks which are between two and three billion years old.

The first cells lived in a water broth, in which the raw materials needed for their reproduction were dissolved. In living higher organisms like ourselves, digestion breaks down the complex molecules in food into usable simpler ones. Our blood is, in some ways, similar in composition to the primordial broth and every body cell is maintained by it. Over the ages, the composition of the sea, like that of the atmosphere, has changed so that our blood is no longer like that of seawater today but instead is much closer in constitution to the seawater of Precambrian times, three to four billion years ago.

The Sexual Revolution. This took place several billion years ago, in a few single-celled plants (algae) and it put the possibilities of evolution on an entirely new footing (Figure 2.3). As long as reproduction was asexual, that is, taking place by simple division of cells to form two replicas of the parent cell, there was little variety for evolution to operate on, and variety is the essential material of evolution. With the appearance of sexual reproduction, the combination of whatever two parents could provide became possible and this led to two huge advantages. The combinations of heritable traits for individuals became almost infinite, enabling evolution to pick

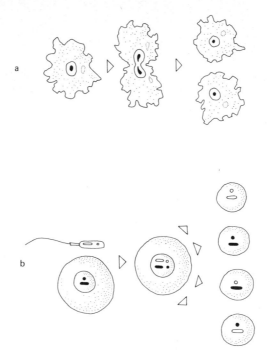

2.3   Sexual and asexual reproduction. A single-celled amoeba (a) splitting in two is the simplest form of asexual reproduction. There is little chance of the daughter cells not being identical to the parent. Sexual reproduction (b) introduces a new genetic material since it depends on the union of two different and unrelated cells. Note that germ cells (egg and sperm) have exactly half the number of chromosomes of the future body cells in the new organism. When this organism produces its own germ cells, the number of chromosomes will again be halved and each may have genetic material to pass on from both parents (as shown by the F_1, or filial one, germ cells in the figure). Thus there is a continued mixing of genes between the sexes and individuals from generation to generation.

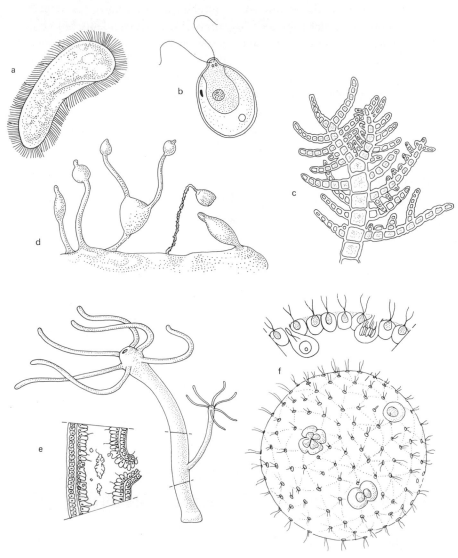

2.4 Evolution: from one cell to many. The earliest steps in evolution were from self-contained single-cell forms such as the animal form *Paramecium* (a), and the unicellular green algae *Chlamydromonas* (b), to cell colonies such as algae (c), slime molds (d), the hydra (e), and the volvox (f). Both the volvox and slime molds are aggregates of independent cells, but nevertheless, develop definite specialized cells for reproduction. The hydra represents an even more sophisticated level of cellular organization with two layers of specialized cells—an outer protective one and an inner one primarily for digestion. A hydra's mouth and tentacles are true organs. The animal reacts to stimuli through special nerve cells. The hydra reproduces sexually but produces buds capable of growing into new hydras asexually.

and choose by natural selection. And beneficial new changes, taking place occasionally, were not confined to a single line of descendants but could be exchanged with other lines, so to speak, so that two different beneficial changes, or mutations, could eventually be combined in the same individuals, greatly speeding progressive evolution. All this will be detailed later, but the meaning of sexual reproduction must be proclaimed here, among the major steps of life on earth.

From One Cell to Many. The next major step in evolution was for one-celled organisms to form more complex organisms consisting of many cells (Figure 2.4). This step took place independently, and by different mechanisms, in fungi, plants, and animals. In algae, the simplest plants, and in unicellular animals, the cell became compartmentalized. In the cellular slime mold (a kind of fungus), many cells joined together and acted as a single organism.

This step toward multicellularity allowed organisms to reach larger size and to perform more efficiently. Specialized tissues and organs could develop for each function, such as locomotion, digestion, elimination, and reproduction, all of which the single cell had to carry out by itself.

Multicellularity introduces two new phenomena into evolution: embryonic development and death. Since the fertilized egg of multicellular organisms is a single cell, an organism must become differentiated into various tissues and organs, and it must grow in size to become like its parents. Since only the reproductive cells, being cells each with half the needed genetic material, can fuse to start the next generation, all other cells in the body—that is, the body as a whole—must constantly reproduce and die.

Under favorable conditions, unicellular organisms never die, nor even fade away. With some exceptions, they reproduce not sexually but by the division of the whole body. The old individual becomes transformed into the new ones, leaving no dead body at all. Those protozoa which have passed up the chance to reproduce sexually live longer than the biblical Methuselah.

Vertebrate Patterns
of Progress

After several billion years, by the time fossils are continuous in the rocks, a number of differently organized major groups of animals had appeared, called phyla. These put basic biological ideas and inventions together in different ways, like worms, or clams, or insects. The last in particular are enormously successful animals; but we call all these phyla invertebrates. We call our own proud phylum the vertebrates because of its vertebral column, or backbone. The basic pattern was a particularly favorable one for evolution to build on, and build it did (Figures 2.5 and 2.6).

The Vertebrates. The oldest identifiable fossil vertebrates are fish without jaws which are recovered from rocks formed some 440 million years ago. They were basically a food straining basket propelled by a tail. Their heads were armored with bone, and their tails were stiffened by a

Figure 2.5

Estimated Time Since the Beginning	Duration (millions of years)	Eras	Periods	Epochs
0.025			Quaternary	Recent
1				Pleistocene
12				Pliocene
28	75	Cenozoic		Miocene
39			Tertiary	Oligocene
58				Eocene
75				Paleocene
135			Cretaceous	
165	130	Mesozoic	Jurassic	
205			Triassic	
230			Permian	
255			Pennsylvanian	
280			Mississippian	
325	300	Paleozoic	Devonian	
360			Silurian	
425			Ordovician	
505			Cambrian	
2000		Pre-Cambrian	Divisions not well established	
		Unknown ages before the formation of rocks now exposed in crust of earth		

2.5 Geologic time scale. (Redrawn from Simpson, George Gaylord. 1961. *The meaning of evolution*, 3d ed. New Haven: Yale University Press, p. 12.)

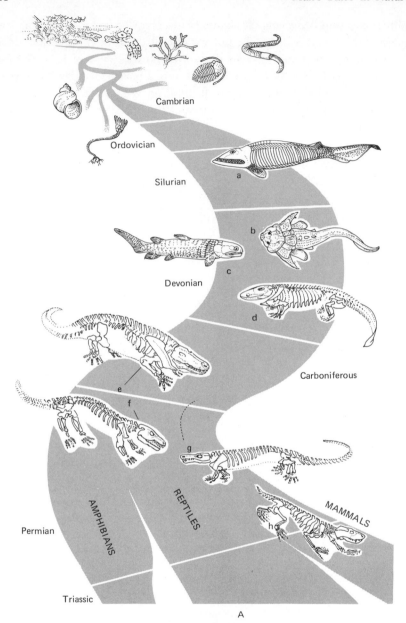

Cambrian

Ordovician

Silurian

a

b

c

Devonian

d

Carboniferous

e

f

g

AMPHIBIANS

REPTILES

MAMMALS

Permian

h

Triassic

A

2.6A *The evolution of life (I)—The Precambrian and Palaeozoic eras:* Geologic time is divided broadly into eras which in turn are subdivided into periods and epochs. Over three-quarters of the age of the earth was passed during the Precambrian era in the distant recesses of which single-cell life was framed from complex molecules of proteins, fats, and carbohydrates (which in turn had been formed from simple atoms and molecules). From single-cell life, complex marine forms evolved and, by Cambrian times (the first period of the Palaeozoic), six hundred million years ago most marine invertebrate phyla had appeared.

rodlike structure (the notochord) made of cartilage, which was the fore-runner of the true bony spine.

Modification of the first gill supports into jaws permitted fish to move from filter feeding to active predation and biting. One group of carnivorous jawed fishes, the Crossopterygii, gave rise to amphibians and, through them, to all four-limbed creatures, including reptiles and mammals.

Two important features permitted the move to life on land. First, crossopterygian fins (actually lobes of flesh) had an axis of relatively large and strong bones. Second, these fish had lungs and could breathe air directly. Why did some fish have leglike lobe fins and lungs? The answer appears to be—so that they could stay in water. Four hundred million years ago, when crossopterygians were living in fresh-water ponds and lakes, there were periods of drought. When one pool of water started to dry up, these fish were able to make their way on land in search of damper places. Several species of lungfish in Africa, South America, and Australia do the same today.

The first amphibians, ancestors to frogs, toads, and salamanders, lived about 375 million years ago. Although more terrestrial in structure than fishes, they were still tied to water, outside of which their eggs and larvae could not live. Furthermore, there was as yet not much suitable food to be had on land. Although terrestrial plants had gotten underway, insects had not to any great extent. So the first amphibians stayed largely in the water and ate fish.

It was the development of a self-contained, waterproof egg roughly 325 million years ago which freed the first reptiles from the aquatic environment. Animals could now move about on land, because their body

---

Primitive vertebrates similar to lampreys arose in the Ordovician (the succeeding Palaeozoic period) but, due to their soft bodies, they are not preserved in the fossil record. The earliest vertebrates that have been preserved are the jawless fishes (a).

Over four hundred million years passed between jawless fishes and the emergence of man. The drawing presents a small fraction of some representative fossil forms along the way.

(a) *Hemicyclaspis*, a characteristic jawless fish.
(b) *Lunaspis*, an armored fish of the Devonian period.
(c) *Osteolepis*, a Middle Devonian crossopterygian.
(d) *Ichtyostega*, the oldest known amphibian skeleton of the late Devonian.
(e) *Eryops*, an early amphibian about five feet long.
(f) *Seymouria*, an early Permian form on the dividing line between amphibians and reptiles.
(g) *Limnoscelis*, a large primitive reptile about five feet long.
(h) *Thrinaxodon*, a mammallike reptile of the late Permian.

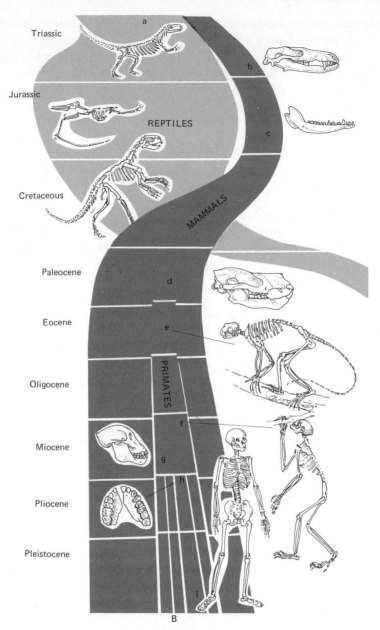

2.6B *The Evolution of man (II)—The Mesozoic and Cenozoic eras:* During the Mesozoic (Triassic, Jurassic, and Cretaceous periods) reptiles, particularly dinosaurs, flourished. Illustrated are *Eurparkeria,* a lower Triassic forerunner of the dinosaurs, *Pterodactylus,* a Jurassic flying reptile, and *Psittacosaurus,* a beaked cretaceous dinosaur.

It is a paradox that, although the replacement of reptiles by mammals took place

fluids also were protected by a more waterproof skin. The so-called Age of
Reptiles was dominated by large dinosaurs (which film-makers alone believe inhabited the earth together with man).

During the Age of Reptiles, mammals were also present but they
were very small and unobtrusive. The first mammals appeared from their
reptilian ancestors approximately 200 million years ago; and while it has
been suggested that they contributed to the extinction of the giant reptiles
by finding and eating their exposed eggs, the diet of the earliest mammals
was mainly insects. In any case, the dinosaurs and other large reptiles survived for another 130 million years, during which mammals remained
primitive.

Primates appear to have arisen directly from these early mammals as
long ago as 70 million years. Man, a comparatively recent arrival, first
began to differentiate himself from other primates something between 10
and 15 million years ago but began to approach his modern size and form
only within the last 2 million.

The Mammals. Besides mobility and intelligence, mammals have
many characteristics which today distinguish them from the other classes
of vertebrates. They are warm-blooded (as are birds), maintaining their internal environment constant within a very narrow range of temperature
and also of biochemical composition. (This trait seems to have evolved in
the extinct dinosaurs and their relations as well.) A modern reptile's temperature varies widely in response to that outside, but a man's tempera-

---

rather late, the mammallike reptiles from which they evolved were among the earliest of the reptilian groups to appear and, in fact, were in a decline before the first
dinosaur appeared on earth.

Mammals themselves are thought to have come into existence toward the end of
the Triassic, but little is known about their history throughout the Mesozoic. Due
to their small size (an average no larger than a rat or a mouse) remains are comparatively rare, limited primarily to tooth, jaw, and small fragments. Illustrated are (a)
Sinoconodon, a triconodont, from the late Triassic and (b) the jaw of *Spalaco-
therium*, (c) a symmetrodont from the Jurassic.

Sometime in the upper Cretaceous a stock of placental mammals appeared from
which more than twenty orders radiated during the succeeding Cenozoic. One of
these, the order of primates, includes monkeys, apes, and man. Shown here are
(d) *Plesiadapis*, a specialized rodentlike primate of the Paleocene, (e) *Smilodectes*,
a lemurlike prosimian of the Eocene, (f) *Pliopithecus*, and (g) *Proconsul*, both an-
thropoids of the Miocene, and (h) *Ramapithecus*, a Pliocene member of the hom-
inid lineage (represented here by separete jaw fragments from India and Africa).
Man (the genus *Homo*) is a comparatively late arrival, although recent discoveries
in Africa have pushed his dates back to perhaps three million. Man of our form,
Homo sapiens is very late, having appeared in Europe not more than thirty-five
thousand years or so ago.

ture is a relatively constant 98.6° Fahrenheit (F); in fact, he cannot tolerate for any length of time more than a 20° variation, that is, a body
temperature below 85° F or above 105° F. This constancy of the internal
environment, or homeostasis, is maintained by a complex neuroendocrine
mechanism. While the mammalian body is covered with fur or hair for
warmth, body temperature is regulated by a center in the brain that controls various mechanisms: blood flow to the skin (Figure 2.7) and sweat
glands, which allow heat loss by evaporation of the perspiration.

Immunologically, mammals are individualists. In man, only grafts between identical twins will take, so highly individualized have we become.
This is what causes so much trouble with heart and kidney transplants.[1]

The central nervous system of mammals is much more highly developed than among reptiles, so that the mammalian responses to stimuli
from the environment are more efficient. The mammalian brain is larger
than that of reptiles (Figure 2.8), better supplied with blood vessels, and
with a blood supply more constant in composition. This is accomplished
by a four-chambered heart (Figure 2.9), rather than the two-chambered
heart which most reptiles possess. The four-chambered heart gives blood
circulation a figure-eight pattern: the heart receives from the tissues venous blood, depleted of its oxygen, sends it to the lungs, where it receives
oxygen, returns it to the heart, whence it flows to the tissues, including
the brain, as arterial blood. Arterial blood and venous blood are, at all
times, separate and unmixed.

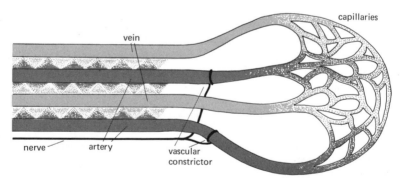

2.7   Heat exchanges between arteries and veins help keep the temperature at the
extremities constant. Arterial blood coming from the heart warms the venous
blood returning and at the same time is cooled. Heat is kept in the body which is
an advantage when extremities are exposed to low temperatures. (Redrawn from
Irving, Laurence. 1966. Adaptations to cold. *Scientific American.* 214:100.)

1. Complex techniques for defeating transplant or graft rejection have been developed,
but only by greatly supressing the normal immunological abilities of the recipient.

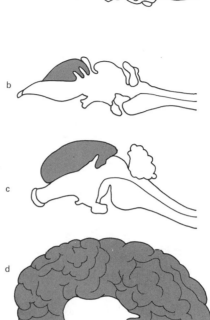

2.8 Cross-section of the brain: (a) fish, (b) reptile, (c) rabbit, and (d) man. Note the difference in the size of the central hemispheres (the higher brain centers) in relation to the other parts of the brain (not to scale). (Redrawn from Magoun, H. W., in Sol Tax, ed. 1960. *The Evolution of Man.* Chicago: University of Chicago Press, p. 200.)

Reproductively, the placental mammals [2] have developed several distinctive organs that slow development and prolong infancy and youth (Figure 2.10); this in turn has a profound effect on their social behavior. The young develop slowly inside the uterus, nourished by a special organ, the placenta, which attaches to the uterus and the developing fetus. The young, born alive but immature, are tied to the mother by the need to suckle milk from the mammary gland (whence the name mammal). Lactation is the basis for the firm bond between mother and offspring. Birds have such a bond, to be sure, without lactation, but the distension of the milk gland makes the mother dependent upon the young in addition to the young being dependent upon the mother; this is not true of most birds.

2. Marsupial mammals (kangaroos, opossums, etc.) do not develop a placenta but nourish their fetuslike young in a pouch.

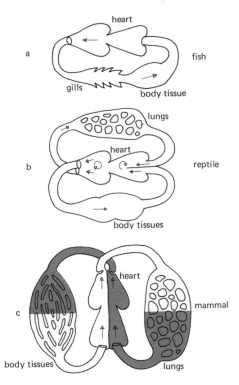

2.9 The heart from fish to mammal showing the transformation of the circulatory system from a single to a double circuit. In the fishes, blood is pumped by the heart through the gills where it receives oxygen, then to the body tissues, and back to the heart. As the primitive lung evolved in air-using fishes and amphibians, gills became lost. Two circulations came into being and, with them, the problem of keeping separate oxygen-rich and oxygen-poor blood in the heart. In amphibians, a three-chambered heart still allowed the blood to mix, however, skin and mouth tissue supplemented the oxygenating process. The four-chambered reptilian heart was a more efficient improvement. Although the partitions were seldom complete, there was little mixing. In mammals, in whom the lungs do all the work, the heart is fully partitioned in four chambers, and no mixing of the two streams of blood passing through takes place.

The emotional bond between mother and child provides a major basis for mammalian society. In many mammals, the sexual bond between male and female provides another basis, particularly when, as it is the way with man, it is permanent rather than transient. The presence of an adult male also enhances the survival of the mother and her offspring. Consequently, young mammals of some species grow up in families which consist of parents and littermates; many others form permanent groups protective of mothers and young. Also, a result of the slow development of mammals, families will often include older offspring of the same parents. This prolonged period of socialized development permits learning and play behavior. Play, a form of learning, leads to exploration and innovation.

A final characteristic which mammals share with birds is the sharp line between waking and sleeping states.

The Primates. Primates are mammals which are particularly adapted

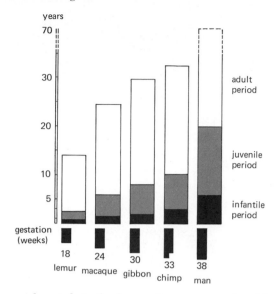

2.10 Developmental periods in the lemur, the macaque, the gibbon, the chimpanzee, and man. (Redrawn from Schultz, Adolph H. 1969. *The life of primates.* New York: Universe Books, p. 149. British and Canadian rights granted by Weidenfeld and Nicolson Limited, London.)

to trees. Although many monkeys (e.g., baboons, macaques) are terrestrial, they basically play or take refuge in trees. Even the large forms that have returned to the ground, such as the chimpanzee, the gorilla, and man show clear traces of their arboreal ancestry. Primates have developed flat nails instead of claws, and hands with flexible fingers with ridged pads at the ends in order to grasp branches firmly (figure 2.11). The primate forearm can be rotated on the upper arm for grasping in any position (in contrast to the more purely hingelike elbow joint typical in quadrupedal mammals). Primates also typically have well-developed clavicles, or collarbones, serving as struts to keep the arms away from the body for extended reach (Figure 2.12).

For life in the trees, vision is more important than smell, and primates have especially good binocular vision (which they share with birds of prey such as owls and the predatory cat family). Their eyes look forward (Figure 2.13), instead of laterally, and have overlapping fields of vision. This stereoscopic vision permits primates to locate objects easily and efficiently in three-dimensional space. Color vision is also well developed.

Touch, likewise, is an important sense and the hairless fingers and toes of primates are covered with many sensitive nerve endings. To accommodate the increased sensory inputs of sight and touch, the cerebral cortex, or outer layer of the brain, which coordinates the movements of

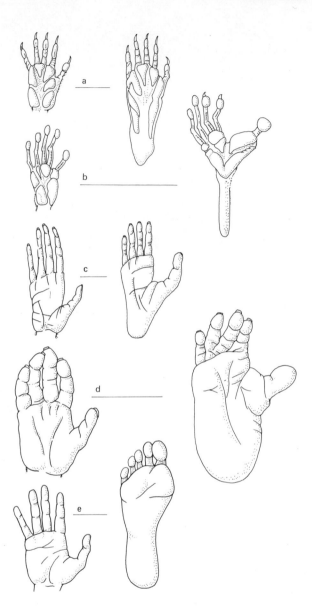

2.11 Hands and feet of Prosimians and Hominoids. (a) *Tupaia* (not a true primate but a good model of the original ancestor), (b) *Tarsius*, (c) *Symphalangus*, (d) *Gorilla*, (e) *Homo*. (Redrawn after Biegert, J. in Washburn, Sherwood L. ed. 1963. *Classification and human evolution*. Chicago: Aldine, 136, 137, 140, 141.)

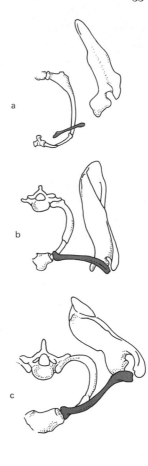

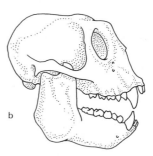

2.12 A comparison of collar bones (clavicles) in (a) cat (in which it is almost vestigial), (b) New World monkey, and (c) man. The shoulder blade anchor of the arm is at the right. The clavicle itself is anchored at the breastbone.

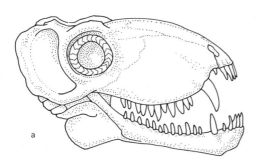

2.13 The development of frontal, binocular vision. (a) The mammallike reptile *Phthinosuchus* and (b) *Mesopithecus*, a Pliocene Old World Monkey. (Both redrawn from Romer, Alfred S. 1967. *Vertebrate paleontology*. Chicago: University of Chicago Press.)

the eyes and hands and provides associations between the various senses, has become greatly expanded (see Figure 2.8 above).

Two biochemical characteristics of the primates are the inability to manufacture vitamin C and the loss of the enzyme uricase. Most mammals other than primates can manufacture their own vitamin C; the guinea pig is an exception. Monkeys, apes, and man, however, must obtain vitamin C from their food. Kittens, puppies, and calves do not have to be urged to drink their orange juice, but we do. If we go without citrus fruits or do not obtain our vitamin C from cabbage, turnips, or other sources for a few months, we develop scurvy. Persons with scurvy have anemia, weakness, painful hemorrhages all over the body, and bone pains. Children fracture their bones, which separate at the ends, where growth takes place. Before the days of canned or bottled fruit juice, sailors used to develop scurvy on long sea voyages. Following the earlier demonstration by James Lind, a British naval surgeon, in 1747, Captain James Cook on his famous Pacific explorations confirmed the value of citrus fruit in preventing scurvy. Lime juice became a standard ration in the British Navy, along with grog, or rum. This is the origin of the term *lime-juicer*, later shortened to *limey*, an impolite term applied to Britons in general.

Vitamin C, called ascorbic acid or antiscurvy acid, is clearly a good thing. So why did the primates lose their ability to synthesize it? It is difficult to see any evolutionary advantage, although the loss seems to have done them very little harm. If they remain in the tropical habitats where they originated, they obtain enough vitamin C from their normal foods. Eskimos may develop scurvy during the long arctic winters, but they make up for it by eating berries in the spring and summer.

The second biochemical loss was the ability to metabolize uric acid, an end product of purine metabolism. Among the mammals, only the great apes and man (and the Dalmatian dog) have lost the liver enzyme uricase which breaks down uric acid to simpler compounds. So only these animals are subject to gout, a painful disease of the joints which is associated with raised levels of uric acid in the blood and deposits of urate crystals in the joints. Fortunately, gout is a rare disease, despite its associations with royalty and the well-to-do.

Why have man and the great apes lost the ability to degrade uric acid? Nobody knows. It is of some interest, though, that uric acid has been reported to be associated with intelligence and achievement. In one study, Army draftees with elevated uric acid levels in the blood scored slightly higher on intelligence tests than those with low levels. And in another study, at the University of Michigan, the more successful and productive faculty members had higher levels of uric acid than the less successful.

The Human Pattern. Physically, man is no longer even partly arboreal, and stands more or less upright (Figure 2.14). This has had a profound effect on his anatomy. Along with other primates who live partly on the ground, such as chimpanzees and gorillas, he is larger than those which have remained completely arboreal. He has developed a unique foot, specialized for walking rather than grasping. His legs are long relative to his trunk (unlike apes which have short legs but long arms) and his thumb easily touches his other fingertips. This is referred to as opposability and greatly enhances manual dexterity. When viewed from the side, man's spinal column has developed curves, insuring flexibility and support for the skull. The skull itself is balanced directly over the spine (Figure 2.15), instead of being far forward, as in monkeys.

The order of development of man's distinctive physical traits was (roughly): the foot, the hand, and the brain. First, the development of the foot permitted an efficient upright posture. Then, the hands, freed from the necessity of grasping, could specialize in manipulation. Finally, the brain increased in average size, from about 700 to 1400 cubic centimeters in less than two million years, a short time as evolution goes.

Man's most important development, of course, is his large brain. It is not the largest in absolute size, since whales and elephants have larger ones. Neither is it the largest brain relative to body size. The human brain weight is about 2 percent of body weight; porpoises and small monkeys outdo us in this ratio. But man specializes in the cerebral cortex, especially the association areas of the brain. Man has the largest proportion of cerebrum to total brain of any animal, the most convoluted cerebrum, and the largest proportion of frontal lobe within that area. This lobe exercises the functions of foresight, inhibition, and worry. In man, the frontal lobe comprises approximately 47 percent of the cerebrum compared with 33 percent for an ape and only 25 percent for that simple primate, the lemur.

The human brain has developed, far beyond any other animal, the interesting property of lateral dominance. Perhaps you are aware that the speech center is usually in the left half of the brain in right-handed people and in the right side in left-handed people. While chimpanzees seem to be right- and left-handed in equal proportion, the tendency toward right-handedness in our own species goes far back in our ancestry. Our predecessor, *Home erectus,* who inhabited the planet some one million years ago seems to have been predominantly right-handed, as was the later Neanderthal man, and all human groups today. There is increasing evidence that each half of the brain specializes in specific abilities as well as emotions. The visual recognition of patterns, the ability to listen to and repeat a series of numbers or musical tunes is markedly different between the two halves of the brain.

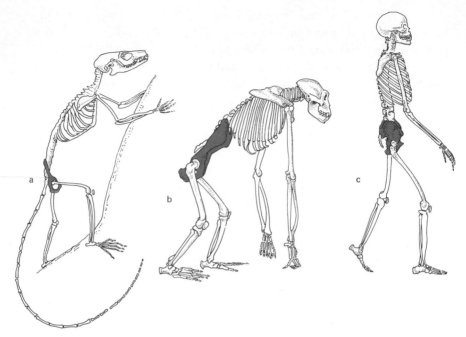

2.14 Orientation of the skeleton and pelvis in (a) tree shrew, (b) gorilla, and (c) man. (Redrawn from Napier, John. 1967. The antiquity of human walking. *Scientific American*. 216:63. And from Clark, W. E. Le Gros. 1963. *The antecedents of man*. New York: Harper and Row, p. 175. British and Canadian rights granted by Edinburgh University Press, Edinburgh.)

The pressure behind the increase in brain size was probably due not so much to the ability to make tools as to the need for an efficient system of communication, that is, speech. Human speech is distinctive, involving unique features both of the brain and of the speech organs themselves. As we have seen, language is profoundly important for human behavior, learning, socialization, and the development of man's distinct product, culture. Human children have a strong drive to mimic the speech of others, and to experiment with language. Apes have now shown that they share some of these logical rudiments of language, but they lack the vocal capacity to experiment with it. Apes communicate emotional states vocally, but are almost incapable of vocal mimicry, because of anatomical differences in the vocal tract. Perhaps while becoming erect, changes in the vocal tract allowed man to form a wider range of vocal sounds, the basis of speech. In any case, as a result, no trained ape has ever acquired more than a few simple spoken words.

Wild apes do, however, have an extensive repertoire of gestures and body language, and they have a large capacity for gestural mimicry. When

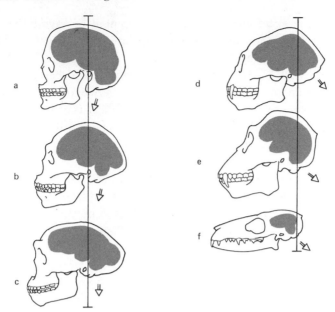

2.15 Primate skulls aligned on occipital condyles (point on which the skull pivots on the spine). Note the expansion of the brain and the reduction of the jaw through the various forms. (a) Modern man; (b) Neanderthal man; (c) *Homo erectus;* (d) *Hylobates;* (e) *Cercopithecus;* (f) *Tupaia* (not to scale). (Redrawn from DuBrul, E. L., and Sicher, H. The adaptive chin. Homans, John, *A textbook of surgery, 6th ed.,* 1945. Springfield, Ill.: Charles C. Thomas.)

experimenters began to exploit this capacity with captive chimpanzees, the animals turned out to be much brighter than had been supposed. One named Washoe learned a sizable vocabulary of signs in the standard American sign language for the deaf (as compared with the maximum of three spoken words learned after great struggle by any previous ape). Other young chimps have been taught to place plastic symbols on a magnetic board to form sentences. All this has opened a window on chimpanzee mental processes, showing a surprising grasp of elementary principles of grammar and syntax.

In finer anatomic aspects, the size of our jaws and teeth has been reduced, our dental arch shortened, and most of our facial and body hair removed. It is not clear what advantage we have gained by losing our body hair, other than an esthetic one. Darwin, in fact, thought that our smooth skin represented social or sexual selection rather than natural selection—that is, man, rather than nature had done the selecting; but the end result is the same: selection favors the survival and reproduction of the favored individuals.

We can sum up man's physical development by saying that he remains a generalized mammal. He can run, climb, jump, swim, and throw. He has avoided the dead-end specialization of being able to do only one of these things well. He has also retained some infantile characteristics: his brain is large and his face and jaws are small, relative to his total bulk. Behaviorally, man grows slowly, lives a long time, and is constantly exposed to others in family and social groupings. He has no mating season, so that sexual interest and activity continue all year round. This has had a profound influence on the human mode of life.

Most kinds of animals are smaller than we are. Of the giant forms of animal life, we are by far the most numerous. In fact, the total human biomass—that is, our average weight multiplied by the total number of people—is the largest of any single species of animal. This can be a comforting or disquieting thought depending on one's outlook. At our current rate of increase, which is far faster than that of other large forms of life, we threaten to crowd everyone off the earth, including ourselves.

Man is geographically the most widely scattered species. He is able to survive at all latitudes and at all altitudes, even under the sea and on the moon. The only forms which approach man in adaptability are those creatures he has carried with him such as rats, lice, and a host of parasites. Through the mechanisms of culture, man has not only adapted to a wide variety of natural environments but has begun to create a uniform artificial environment for himself around the world. Americans wintering in the Antarctic, or summering in the tropics, live under conditions not too different from those back home. They can raise the heat or turn on the air conditioning, watch television, go to the supermarket, and listen to Muzak. Of course, not all of the elements of this truly remarkable environment are pluses. Along with cars there are traffic jams. Beautiful beaches are strewn with beer cans, and even such erstwhile Edens as Hawaii, Tahiti, Fiji, or Borneo have not been spared the onslaught of rock music and plastic rubbish.

SUGGESTED READINGS

Campbell, Bernard. 1967. *Human evolution.* Chicago: Aldine.
Clark, W. E. Le Gros. 1963. *The antecedents of man.* New York: Harper and Row.
Harrison, Richard J., and Montagna, William. 1969. *Man.* New York: Appleton Century Crofts.
Howells, William W. 1973. *Evolution of the genus* Homo. Reading, Mass.: Addison-Wesley.

Huxley, T. H. 1863. *Evidence as to man's place in nature.* Chicago: William E. Yorgate.

Pilbeam, David. 1972. *The ascent of man.* New York: Macmillan.

Washburn, S. L., and Jay, P. C. 1968. *Perspectives of human evolution.* New York: Holt, Rinehart and Winston.

3
EVOLUTION AS
A PROCESS

That evolution is a fact needs no belaboring today, although Darwinians a century ago were very busy defending his ideas. It is not necessary to repeat all the arguments, but one approach, which I shall follow in this chapter, is to see things as Darwin saw them. That is, to study the adaptations of animal form and the various signs of past evolution they give; to pay particular attention to Darwin's basic contribution, natural selection, while noting other mechanisms which have been added to fill out the picture; and to begin by pointing to the various kinds of evidence of evolution which he used together with others used later on.

Nature's Witnesses

Geography. Geographic distribution of various species reflects the fact that each form could have origniated at only one place and at only one time. The discontinuous distribution of related forms indicates that they migrated to their present residences and became extinct in intermediate areas. Thus kangaroos are found only in Australia and New Guinea, where they evolved. But the larger group to which they belong, the marsupials, occurs in Australia and the Americas, while fossil forms are found in between. Again, islands in the South Atlantic midway between South America and Africa have plants that are indigenous to both these continents.

Comparative Anatomy. There are remarkably similar configurations in the anatomies of related animals. Almost all mammals have seven neck vertebrae and all have four-chambered hearts. To quote Darwin, "What can be more curious than that the hand of man, formed for grasping; that of a mole, for digging; the leg of a horse, the paddle of the porpoise, and the wing of a bat should all be constructed of the same pattern and should include similar bones in the same related position?" (See Figure 3.1.) Fur-

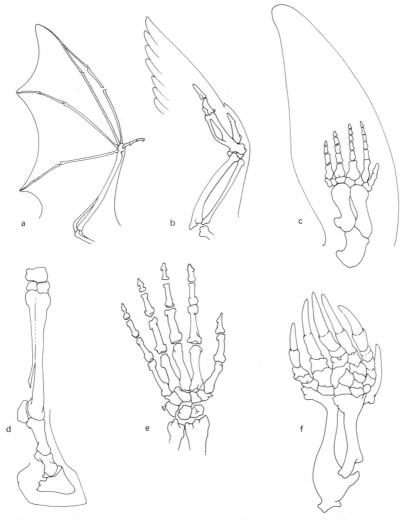

3.1 The similar physical structure of widely divergent forms convinced Darwin that vertebrate species were not created separately but evolved from a common ancestor. Note the homologies; for example, all have two bones in the "lower arm" whether they are used in rotating the limb or not. (a) Wing of a bird; (b) wing of a bat; (c) flipper of a whale; (d) leg of a horse; (e) hand of man, (f) forelimb of a mole.

ther examples need not be given, but it is striking that such similarities are closer in the earliest stages of the embryo than at any later stage. In fact, it is almost impossible to distinguish by inspection human embryos at four weeks of gestation from those of lizards, birds, or other mammals at an equivalent stage of development.

Another aspect of comparative anatomy in the study of evolutionary relationships is the presence of vestigial organs, such as the minute pelvis and skeletal hind limbs of the python or the unerupted teeth of baleen whales (which strain their microscopic food) or the vestigial wings of flightless birds such as the ostrich.

Biochemistry. While the most direct method of study has traditionally been that of fossil remains, a recent development lies in the study of proteins of related animals. Protein molecules are made up of long chains of amino acids, hundreds to a chain. By counting the numbers of substitutions of one amino acid for another, in the same protein from one species to another, one obtains a numerical expression of their relationship. The fewer the differences, the closer the relationship. Unfortunately, since the rate of substitution is not constant but differs for different amino acids, the construction of accurate time scales of these substitutions in the evolutionary past of related animals is still under intensive investigation.

There are also other biochemical tests for the relatedness of animals. One is the amount of precipitation produced by mixing the blood serum of two species. The more closely the two animals are related, the less precipitation there will be when their sera are mixed. By this test, as by the analysis of proteins, human blood is less distant from that of the great apes than from that of monkeys, lower primates, and other mammals in that order.

Chromosomes. Another recent development is to compare the number, size, and shape of chromosomes in the cell nucleus, constituting the vehicle of heredity. Man has twenty-three pairs, while orangutans, chimpanzees, and gorillas have twenty-four pairs. All of these are closer in number to each other than they are to other primates which, in turn, range from seventeen to thirty-six pairs. The amount of chromosomal material, however, is more constant than this range of numbers would suggest since when they are many in number, they tend to be small in size. In the past few years, it has become possible to identify chromosomes individually by improved staining methods. Thus human and other primate chromosomes can now be compared in fine detail and this has shown that man's chromosomes are closest to those of the chimpanzee and gorilla.

Parasites. The study of parasites can also be used to illuminate evolutionary relationships. Parasites are almost as old as life itself; fossil parasites have been found dating to some three hundred million years. These probably had already evolved from previous free-living forms. Many parasites have become so adapted to their hosts that they are restricted to a single host species. Certain lice are found only on man and the chimpanzee; in all probability, the ancestor to these lice lived in the ancestor believed common to both man and chimpanzee (*Proconsul*), perhaps twenty million years ago.

Comparative Behavior. A final method of study is that of comparative behavior. The well-known behaviors of ants, bees, fish, and birds show great similarities within related forms. Observations of the higher primates reveal facial expressions, attitudes, postures, and posings that are disturbingly like our own.

The Process of Evolution

The Spur of Natural Selection. From his long and widespread observations, Darwin saw how well different forms of life were organically fitted, or adapted, each to its habitat or way of life. He also recognized the great importance of variation within a species or variety, variation that had prevailingly been regarded as an accidental or inconsequential straying from the ideal form (such as a species would have in the traditional creationist view). When variations are heritable, he saw, it is possible for the norm of the population to be changed, as breeders had long done with domesticated plants and animals by selecting for breeding the favorable variations they find. Darwin, in his great logical step, transferred this to nature: from the varying members of a generation, nature selects the most fit, the best adapted, to succeed and thus to contribute most of the heredity of the following generation. This principle of natural selection, so well understood now, was the reason for Darwin's explosive impact. In addition, when two populations of a species become progressively adapted in different directions, they diverge in their genetic, or heritable, endowments. When their genetic differences have accumulated to the point where they do not, or cannot, interbreed, they have become two separate species. They have speciated, nevermore to exchange their heritable features.

Variation and Its Importance. Let us consider this in more detail. As suggested above, plants and animals, including man, occur as species,

3.2   Human variation. With a world population of four billion, no two individuals are the same. Man's genetic differences are among his species' most important possessions. (courtesy Monkmeyer Press, photo by B. Anderson)

which are groups of populations able to interbreed mutually. They exist in no other condition. The interbreeding populations are not all alike; they vary genetically (for example as races). The individuals within a population also vary genetically (Figure 3.2), and even the genetic differences within a family are considerable. The differences of individuals—their variation— is of two kinds: measured and counted. Most measurable characteristics of

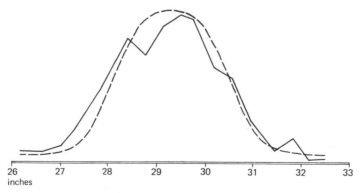

| 26 | 27 | 28 | 29 | 30 | 31 | 32 | 33 |
inches

3.3   The bell-shaped distribution: seated height of American adult males. The dotted line indicates the idealized normal distribution within a large population. The solid line shows the actual distribution (in percentiles) in a study of 309 individuals by author.

a population, such as height, weight, or intelligence, can be represented by the familiar bell-shaped curve (Figure 3.3), which has an average value or mean with a spread (the variation) around it. Tall and short men are as normal as men of average height; they are merely less common. Average characteristics and the range of variation surrounding them will differ from one species to another, and from one population to another within a species.

The other large category of human traits is counted rather than measured. The blood groups are examples (Figure 3.4). These are not measured in terms of how much, but rather what kind. The variations can be counted, and frequencies can be determined for both populations and species.

Part of a population's variability is genetic and part is environmental. All characteristics result from the interplay of heredity and environment. The genetic makeup of an organism, determined by the genes inherited from the parents, is the genotype. What the organism actually looks like at any time is the phenotype. A person has the same genotype throughout life but has different phenotypes, as fetus, child, and adult (Figure 3.5). We shall discuss these points more fully later, along with other genetic details.

The Sources of Change. A variation which appears without any change in the environment and which continues to appear in the offspring, is genotype variation. In evolutionary terms, this is the only kind that is important. It has two sources, mutation and recombination. Either mechanism can affect genes or chromosomes. All sexually reproducing animal species have spontaneous mutation, with different rates for specific genes. Mutations can also be induced artificially by X rays, ultraviolet light, extremes of temperature, and chemicals. Such mutations have been

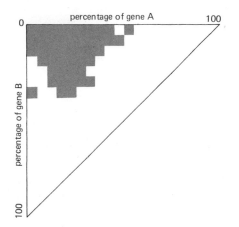

3.4 *ABO* blood group gene frequencies as distributed in 215 populations throughout the world. Since blood groups do not occur at random, there is a concentration of 15 to 30 percent for *A*, 5 to 20 percent for *B*, and *O* varies from 50 to 80 percent. (SOURCE: Brues, A. 1954. *American Journal of Physical Anthropology.* 12: 559–597.)

3.5 While one's *phenotype* changes dramatically through life, one's *genotype* never changes.

unimportant in past evolution, but they may become increasingly important in the future.

The actual nature of gene and chromosomal mutation will be discussed in later chapters. Here it should be noted that the recombination of genetic material already present accounts for far more of the observed variation within a population than does mutation, which is a rare event. But the only way new genetic material is produced is by mutation. Of course, new genetic material can be introduced into a population as a result of mixture with other populations. This is referred to as gene flow

Heritable variations appear constantly in a population. They are then selected, either for (if they are beneficial), or against (if they are harmful). The great majority of mutations, however, are harmful. In either case, natural selection works to eliminate the unfavorable and preserve the favorable. This is done through two mechanisms, differential reproduction and differential mortality.

The favorable mutation may result in more offspring being produced by its happy possessor. To do any good in an evolutionary sense, these children, some or all of whom will have the mutation, must survive to reproduce in their turn. Darwin was so impressed by the prodigious potential fertility of nature that he thought differential mortality was the main agent of natural selection. Modern workers are more inclined to stress differential fertility as more important—that is, possession of the good mutation enhances one's reproductive success.

In either case, the fittest survive. Fitness has nothing to do with health, intelligence, value to society, or accomplishment of any kind other than reproductive. The fittest survive: the more they survive, the fitter they must be considered. From the standpoint of biological evolution, a half-wit with fourteen grown children is more fit than a genius like Leonardo da Vinci who was childless.

Natural selection, whether brought about by differential fertility or by differential mortality, can take three directions: stabilizing, directional, or diversifying. When a population is well adapted to a stable environment, then selection will favor the preservation of the adapted type, that is, the species average, or norm. Stabilizing selection is sometimes called normalizing selection for that reason. But if the environment changes, or if new mutated form is clearly an advantage in a constant environment, the norm of the population will shift toward the new type.

Diversifying selection is favored when there are two or more adaptive backgrounds which can be occupied by populations of a species (Figures 3.6 and 3.7). It may also result from other advantages to a population arising from genetic diversity—response to different seasons, flexibility in responding to varied stresses, and so forth.

Both diversifying selection and directional selection can lead to the formation of a new species. Breeding populations, which are what species are composed of, can become isolated in several ways. Geography can interpose physical barriers such as oceans, mountains, and deserts (Figure 3.8). Geography is the most important isolating mechanism. Ecologically the populations may occupy different niches or habitats, or they may breed at different times. Genetically changes in the chromosome mechanism may prevent interbreeding. Behaviorally a host of mechanisms prevents interbreeding of subpopulations or subspecies. Visual or olfactory cues are important as well. If any of these mechanisms leads to breeding isolation, which lasts long enough or becomes complete, the two subspecies will no longer be able to interbreed. They will have become new species, each adapted to its own environment, in the same way, and at the same time.

Drift: The Role of Chance. So far we have mentioned as the mechanisms of evolution heritable variation, natural selection and isolation. We noted that gene flow or migration or mixture between populations will introduce new genes into the population. There is, however, still another set of factors at work called genetic drift—the chance fluctuation in genes from one generation to another. In any small group, accidents like drowning or warfare may by chance wipe out a family, removing all cases of, say, blood group B from a breeding isolate. Over the years, many such

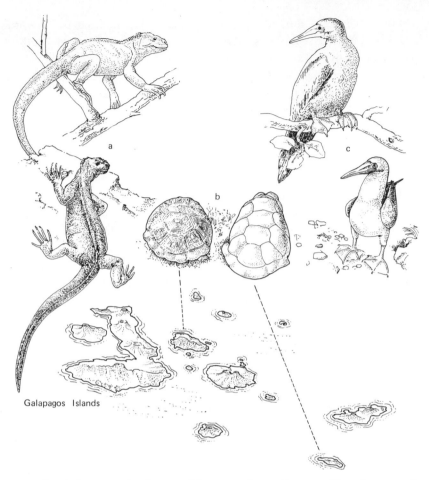

Galapagos Islands

3.6 The Galapagos Islands are a laboratory for evolution. Although outwardly similar in appearance, marine and land iguanas belong to different genera. Through time, they have evolved separately while exploiting very different ecological niches. Similarly, the red-footed booby (c), by nesting in trees, does not compete for space with its blue-footed ground nesting relative (a bird of a different species). Isolated and mutually exclusive gene pools on different islands have produced characteristically different shell shapes on the giant tortoise (again, different species).

events will occur, so that two groups that may have had a common origin will differ considerably, if there has been no gene flow between them. This has been demonstrated by people living in neighboring hostile villages in New Guinea. As a result of chance genetic drift in this area, two groups speaking dialects of the same language will differ markedly in the distribution of their blood groups.

3.7 Some of Darwin's Galapagos finches: (a) vegetarian tree finch, (b) sharp beaked ground finch, (c) large ground finch, (d) small insectivorous tree finch, (e) tool-using finch. In all, Darwin discovered thirteen species on the Galapagos, each of which evolved to exploit a particular ecological niche. One of the most unusual birds anywhere is the tool-using finch, which, because there are no woodpeckers in the Galapagos, fills that particular niche. Because it lacks the woodpecker's long tongue to dislodge insects, it has compensated by learning to use the spine of a cactus.

Although the mainland ancestor from which these diverse birds evolved is unknown, it is thought to have been a small, thick-billed, ground-feeding, seed-eating, finch species of the open country.

A very special case of genetic drift is called the founder effect. A group, usually a family, may decide to leave the main band and establish a separate village. Families tend to have more genes in common than will be found in the general run of the population. Again, by chance, all persons with blood group *B* may be in the new village, leaving none behind. Thus the two groups will differ, although they originally came from the same population.

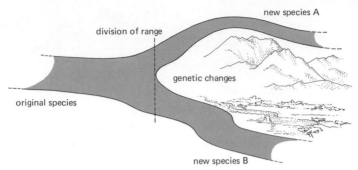

3.8   Speciation brought about by the geographic isolation of two (or more) breeding populations. When a species inhabits new territory, some of the population
may make new and different adaptations. In this or other ways the genotypes
become incompatible, so that the populations do not interbreed normally. At this
point, it is usually said that new species have been formed.

The Pattern
of Evolution

The agents and processes of evolution tend to produce a typical pattern,
which can be summarized. First, there is divergence, as one or more new
species appear from a previous one. Each extends its adaptation to its particular environment. If a species develops a particularly flexible or favorable organ or adaptation, it may expand broadly and rapidly in an adaptive
radiation to occupy a variety of niches or habitats, giving off daughter
species; and in the same situation actual changes or progress may be
rapid, as in the case of the expansion of the human brain, at one or two
points in our evolution.

   With a settled and successful adaptation there will ensue a period of
stability, with little further change. Some living forms are extremely ancient and unchanged, like the horseshoe crab (Figure 3.9). The final
event, extinction, has been the fate of over 95 percent of all species that
ever existed. While some prophets of gloom and doom predict the same
fate for man, most human biologists are hopeful that we shall learn how to
manage our own evolution so this will not occur.

   In the long course of evolution various proteins, organs, and animal
species evolved at different rates, at different times, and at different
places. This means that instead of a smooth progression from one form to
another, with the first conveniently disappearing just as its successor was
getting underway, the two (or more) forms have overlapped. For a long
time, this point was not appreciated in the study of the evolution of man,

3.9 Rates of evolution may differ radically according to environmental changes and pressures. The possum, tuatara, cockroach, and horseshoe crab in column (a) have changed little over millions or even (in the case of the horseshoe crab and the cockroach) hundreds of millions of years. In approximately sixty million years, however, the horse [column (b)] evolved from a foxlike creature of the Eocene to the modern form we know today. Column (c) shows an example of very rapid microevolution. In response to industrial pollution, English moths, whose safety depends on their ability to blend into the background, have changed from light to dark (melanic) forms. As pollution replaced light-colored lichens with soot, melanic forms became favored. In some heavy industrial areas, within only fifty years, dark forms constituted 98 percent of the moth population.

and it was difficult to reconstruct the human family tree from the fossil record. It is now believed that what was essentially modern man (*Homo sapiens*) already had appeared during the time of the Neanderthals, who, in all probability, represented a dead end. Rather than dying out completely, however, they probably contributed some genes to the pool of modern man.

The question is often asked whether fortuitous changes, in the form of random mutation, could ever be the source of the marvelous variety of life which exists in the world. The answer must be yes. With strong selective pressure, evolution, large enough to be measured, can occur in a surprisingly small number of generations. (The famous example of industrial melanism in moths took fifty years (Figure 3.9), or fifty generations, from the appearance of the first black moth in industrial towns in England to a 98 percent frequency. In Australia, where rabbits had become a pest, a virus was introduced in 1950 which killed 99.8 percent at first. By the late 1960s, the mortality rate had fallen to 5 percent, as a consequence of genetic adaptation by both virus and host. In this case the virus had become less lethal and the rabbit considerably more resistant to its effects.

For a human example, the frequency of carriers of the gene for sickle-cell hemoglobin (hemoglobin S) in West Africans brought to America, where there was no malaria, fell from about 20 percent (an estimated value for the tribes they came from) to 9 percent in two hundred years, having been halved in six generations.

So evolution can work fast, and there has indeed been enough time for the evolution of all the variety of life we now see. And, we noted, it is still going on. As we shall see in later chapters, the rules may have changed a little and the referee is less often nature and more often man himself, even though he is hardly conscious of the fact. But the game is still played.

SUGGESTED READINGS

Dobzhansky, T. 1970. *Genetics of the evolutionary process.* New York: Columbia University Press.

Mayr, Ernst. 1963. *Animal species and evolution.* Cambridge, Mass.: Harvard University Press.

Romer, Alfred S. 1967. *Vertebrate paleontology.* 3d ed. Chicago: University of Chicago Press.

Simpson, George Gaylord. 1961. *The meaning of evolution.* 3d ed. New Haven: Yale University Press.

Stebbins, G., and Ledyard, Jr. 1971. *Process of organic evolution.* Englewood Cliffs, N.J.: Prentice-Hall.

4
GENES: THE PHYSICAL BASIS OF HEREDITY
By Jonathan S. Friedlaender

While the great sweep of evolution—the adaptive changes in animal species which occur over time, and even the almost inevitable extinctions—is ordered by natural selection, for a long time it was a puzzle exactly what caused new variants to appear in a given animal or plant species, or even how existing variation in all manner of characteristics was transmitted from parents to offspring.

The Historical Perspective

Darwin's Problem. Darwin saw clearly that such heritable variation within species was a necessity for the process of evolution to work. It is this heritable variation which gives some organisms the edge in surviving and reproducing, so their particular hereditary characteristics should be passed on to more of the following generation than those of others. This elegant line of reasoning had a flaw in the mid-nineteenth century: almost no one had a clear idea of what we now call genetics—how organisms come to inherit their particular constellation of physical characteristics. In a vague way, most scientists thought that maternal and paternal hereditary factors blended together in their offspring, so that in all respects, children sould be averages of their parents.

Darwin's contemporary Fleeming Jenkins (actually an engineer) pointed out that, if this were so, the heritable variation in a species, so crucial to Darwin's argument, would collapse around the averages within a very short time. Even worse, new varieties would be washed out as soon as they appeared. What possible importance, then, could natural selection have? Darwin could never provide a completely satisfying answer, although a contemporary of his in a Moravian monastery had the key.

Mendel's Solution. Father Gregor Mendel, raised as a peasant, a failure as a teacher, quietly discovered some basic truths, in his lonely garden, about the mechanisms of heredity. He cross-fertilized different varieties of peas (short and tall, green and yellow, rough and smooth seeds, and so forth) and found that the offspring were not an average of the parental characters (Figure 4.1). In the case of tall and short plants, all the progeny were as tall as the tall parents. Even more startling, when these tall hybrids were allowed to fertilize one another, their offspring were mostly tall, but some, approximately one out of every three, were short like the original short strain. Mendel reasoned that the original variant traits did not blend in offspring, but maintained their integrity, or particulate nature, in the new combinations. Their particular effects might be masked—one trait might be dominant to the other, not allowing its expression—but the basic units of heredity, or genes as they came to be called later, were unchanged. He saw that in sexually reproducing organisms, these units come in pairs, one from each parent. The constant reshuffling of these units in generation after generation of sexual reproduction is the source of sustenance of variety and of evolutionary possibility.

In Mendel's experiments lay the key to the Darwinian problem. But no one saw it. Mendel's paper, published in an obscure journal, was neglected. In publishing he achieved instant anonymity. Discouraged, he became an administrator. Only much later did scientists understand the importance of his key in unlocking the problems of evolutionary theory.

The Operations of Genes. As for understanding more about the nature of the hereditary units, or genes, by 1902 it was suggested that the dark-staining bodies (or chromosomes) in cells were the physical structures which carried the genes (Figure 4.2). They divided and replicated themselves in the course of cell division and gamete formation, and they occurred in pairs in most organisms. Their activities were compatible with what Mendel had established about the nature of the genes.

Understanding exactly what genes are and how genes work on the biochemical level was a problem of a different magnitude, and it began to come into focus more clearly by about 1940. To recapitulate, the attri-

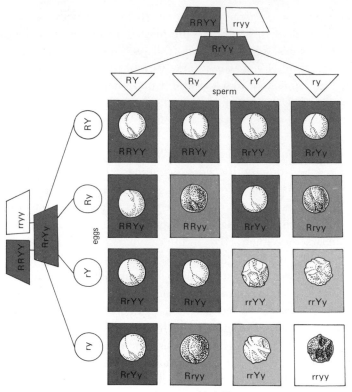

second hybrid generation

4.1   The Punnett Square shown here graphically illustrates Mendel's law of the independent assortment of genes. The square shows two generations. In the first are pure strain parents, one possessing the dominant characteristics of roundness and yellowness (*RR* and *YY*), the other two recessive characteristics of wrinkledness and greenness (*rr* and *yy*). These strains are crossed. The resulting first generation hybrids must combine all the four genes of the parents, i.e., *R*, *r*, *Y*, *y*. Phenotypically the hybrid appears identical to the parent with the dominant traits, that is, round and yellow. However when the first generation hybrids are mated, their available genes, both dominant and recessives, produce the various combinations shown in the square. In the second generation four phenotypically differing kinds of peas are produced in the approximate ratios 9:3:3:1.

Mendel discovered this by keeping careful statistics and finding such ratios. This also illustrates the important fact that all the genes survive. *Dominant* does not mean that dominant genes become more numerous than recessives. This persistence of particulate units of heredity, unknown to Darwin, was the missing piece in his understanding of the vital fact of variation.

butes of the genes were well established. They are discrete and unblending, as Mendel had perceived almost a century ago, so that variation among organisms is not averaged out, or lost, over the generations. Secondly, the structure of the hereditary material has to be variable enough

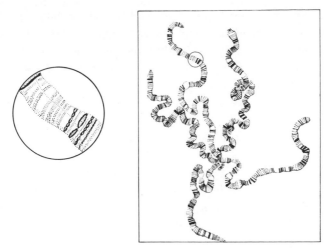

4.2 Salivary gland chromosomes of *Drosophila melanogaster*. The dark bands are believed to be gene loci.

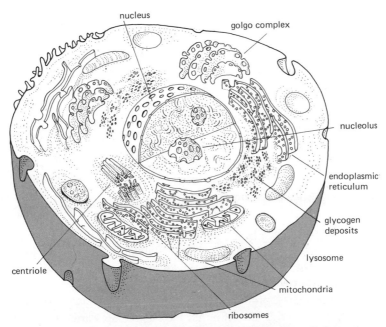

4.3 A schematic cell (from a rat liver). The appearance of a cell depends on the type of tissue from which it comes (muscle, nerve, skin, bone, blood, etc.). All cells, however, have a nucleus (which controls many cell activities and where chromosomes are located). This is surrounded by cytoplasm which contains many important structures, including ribosomes, the sites for protein synthesis. The cytoplasm is enclosed by a membrane, the cell wall. The various labeled structures are shown to demonstrate that even a single cell is highly complex.

to carry or store information, just as the alphabet allows for an almost infinite number of combinations, some of which, appearing on this page, will mean something to a captivated reader. By comparison, invariant structures or patterns can convey next to no information. And not only does there have to be a code in the genetic material, but somehow the code has to be translatable into the control of cellular differentiation and functioning. [Figure 4.3]* This was all the more remarkable because it certainly appeared at that time that the chromosomal makeup of almost all the cells in the body is constant—that hair follicle cells, for example, have the same chromosomal constitution as white blood cells.

Another remarkable attribute of the genes is their obvious ability to replicate themselves, or make copies of themselves, almost free from error. Every time a cell divides genetic material is produced to enable the two daughter cells to be endowed with the gift of a complete set of genes essentially identical to those previously carried by the parent. Of course, the copying system was clearly not perfect; scientists were well aware of the appearance of completely new hereditary forms, or mutations. And clearly, if there were no possibilities for the introduction of new "words" or "sentences" into the language of heredity, evolution would not have got very far.

DNA, the Fundamental Material

Not only was this long list of immensely important and complex functions attributable to the genes, but geneticists knew by the late 1940s that the genetic material, in the chromosomes of all organisms with decent nuclei, had to be a substance now referred to as DNA, or deoxyribonucleic acid (Figure 4.4). While the chemists had not come to grips with the structure of this apparently complex sort of molecule, almost everyone assumed that because of the wide variety of tasks the genes performed, the structure of the genes would prove to be immensely complicated—at least as complex as the structures they produced, the myriad proteins and enzymes that are the stuff of everyday life. After all, life has the appearance of being incomprehensible in its innermost mechanics, so why should anyone expect the hidden instructions for the circuitry to be easily decipherable?

* Throughout the text, when a figure reference is enclosed within brackets, that particular figure is relevant to the general discussion rather than just to the specific passage with which the figure reference appears.

4.4 Segment of the DNA double
helix. In the center are bases of
purine and pyrimidine. The dark
chains of the helix are sugars and
phosphates.

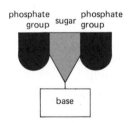

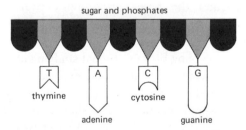

4.5 A nucleotide: The unit of DNA is composed of three parts—a phosphate
group, a sugar, and a base. Bases are either purines or pyrimidines which are
bonded to the sugar. The sugar of one nucleotide is in turn bonded to the phosphate group of the next and so on making a chain.

The purine bases of DNA are adenine (*A*) and guanine (*G*), which are larger and
more complex than the pyrimidine bases, cytosine (*C*) and thymine (*T*).

Discovery of DNA. In fact, there were a number of clues about the nature of DNA. It was known that the large DNA molecules are formed from three sorts of smaller molecules—a sugar (deoxyribose, with five carbon atoms), phosphoric acid, and four varieties of bases (Figure 4.5). Two of the bases, cytosine and thymine, are very small in size (classed as pyrimidines) and the other two, adenine and guanine, are somewhat larger (classed as purines). Each of the four combines in DNA with a molecule of sugar and phosphoric acid to form a weak acid found, of course, in the nucleus, and hence their names, nucleotides. These building blocks alone are the constituents of the enormously long DNA molecules. It was their arrangement that somehow provides for the workings of cells, the creation of new individuals, and carries the instructions for life.

It was also known by the late 1940s that there are substantial regularities in the arrangement of these materials. The phosphate and sugar units were known to alternate along the long strand of DNA. Because of their regularity, they could not carry the code of life. It had to be the sequence of bases that carries the key. Actually, by that time it had also been established that the relative amounts of the four different bases are the same among different cells of the same individual and, in fact, among cells from animals in the same species; yet, in different species, the ratios are different. Despite this, the amount of adenine always equals the amount of thymine, and the amount of guanine always equals the amount of cytosine. Even further, some diffraction studies carried out by Maurice Wilkins and Rosalind Franklin in London indicated that the general structure of the DNA molecule is a spiral, or helix.

It was at this point that Linus Pauling, the great structural chemist, proposed a hypothetical structure for DNA. It had three spirals, or helices, running parallel, with the alternating sugars and phosphates running down the inside and the bases projecting outward with their undeciphered messages. The model was unsatisfying because it gave no clue as to how DNA performed its complicated tasks.

James Watson, a young biology student studying at Cambridge in the early 1950s, and Francis Crick, then thirty-five, had other, more elegant ideas. Watson pointed out that important biological objects come in pairs—not in threes—and genetic material should be no different. He and Crick tried building models of DNA consisting of two strands, with the sugar-phosphate structure running down the outside, and the bases somehow paired in the middle. Very quickly Watson hit on a beautiful model, when he saw that an adenine-thymine pair held together by hydrogen bonds was identical in shape to a guanine-cytosine pair (Figures 4.6–4.8). In each position along the spiral staircase, there was room only for a small

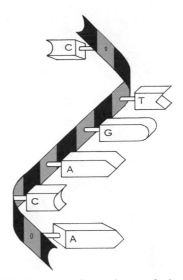

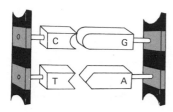

4.6 In DNA, cytosine can bond only with guanine and adenine only with thymine.

4.7 A section of a single strand of DNA.

and large base combined. Not only that, adenine and thymine could be joined by two hydrogen bonds, while cytosine and guanine could be held together by three.

Immediately, they knew they had something so simple it had to be right. If the base-pair-matching arrangement was correct, then one side of the helix could specify what sequence occurred on the other—it could act as a mold, or template. The complementary nature of the bases immediately suggested how the duplication of the genes took place—if the bases came apart down the middle, they each could act as molds specifying their other halves, so that two identical spirals could result (Figure 4.9). The structure they proposed also suggested how other properties of the gene were fulfilled. The information, or code, is in the sequence of bases, carried inside the helix. Because there are only four bases in DNA, and twenty different substances (amino acids) which commonly appear in protein sequences, obviously one base in the DNA does not specify the appearance of one particular amino acid in the synthesis of protein. It had to be a sequence of bases—pairs are not enough, as four things can only be paired in sixteen different ways; therefore the minimum for "word" lengths of bases had to be three. We now know this to be the case, and the code has been successfully deciphered. As for the possibilities for

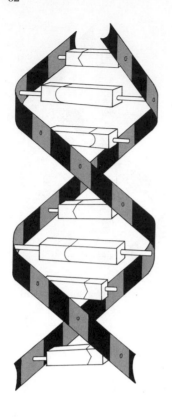

4.8   The double helix—matching segments of strands like those in Figure 4.7. An American, J. D. Watson, and an Englishman, F. H. C. Crick, together working at Cambridge University first proposed this model of the structure of DNA in 1953, and since then, research appears to have demonstrated its reality. Note that it is the bonding of the bases which holds the double helix together.

change in the code, obviously if there is some sort of error in the duplication process, this would become established in all following divisions, unless some editing method excised it.

   Reading the Code. Exactly how the code of bases in the DNA was deciphered in the workings of the cell was mainly elucidated by Crick. The DNA is in the nucleus of the cell, in the chromosomes. Yet the basic workings of the cell, its metabolism, mainly the making of proteins, occur outside the nucleus in the cytoplasm. How is the information passed out and translated?

   As depicted in Figure 4.10, the code of bases along one side of the DNA is transcribed into a single-stranded molecule with alternating bases, called RNA, or ribonucleic acid, indicating that the structure of the sugar is slightly different from DNA. This form of RNA carries the DNA message from the nucleus to the site of cellular metabolism in the cell, called ribosomes, and hence is called messenger-RNA. At the ribosome, the message is translated into the chains of amino acids (called polypeptides) which form the structural stuff of life (Figure 4.11).

   The code is the same for all living organisms. Three adenines in succession in DNA will specify the same amino acid in a bacterial cell as in a human cell. The code was perfected early in the history of life. Note that there are redundancies in the code. Some amino acids can be specified by more than one base combination; always in such cases it is the third base that can be altered without changing the amino acid. There are also three

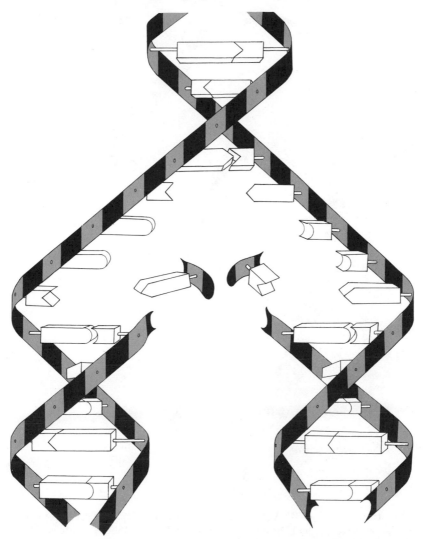

4.9 The replication of DNA. The component strands of the double helix uncoil and separate at the bases. Individual nucleotides, already present in the nucleus find their way to matching positions and thus new strands of DNA are formed on the old. Due to the programmed pairing of the bases this new strand will duplicate the original complement of the helix.

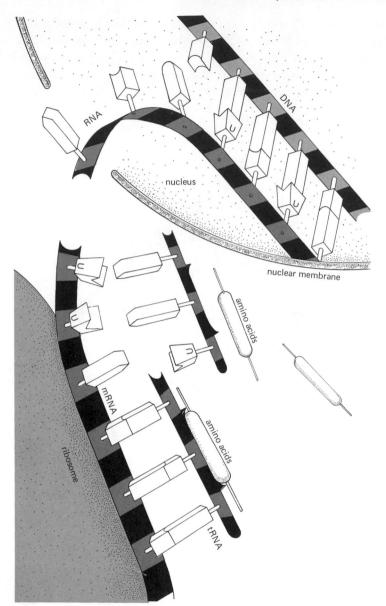

4.10 The synthesis of RNA. RNA differs from DNA only in that thyamine is replaced by uracil (U) and its sugar is slightly different. RNA is synthesized on a single strand of DNA in much the same way as DNA reproduces itself; therefore, it is practically the duplicate of its DNA model. This type of RNA is produced in the nucleus which it leaves through a pore in the nuclear membrane. In the cytoplasm, messenger-RNA (mRNA) in conjunction with ribosomes and transfer-RNA (tRNA formed in the cytoplasm outside the nucleus) bring together amino acids of various types to make a polypeptide chain. tRNA assembles in triplets and each set attracts a different amino acid.

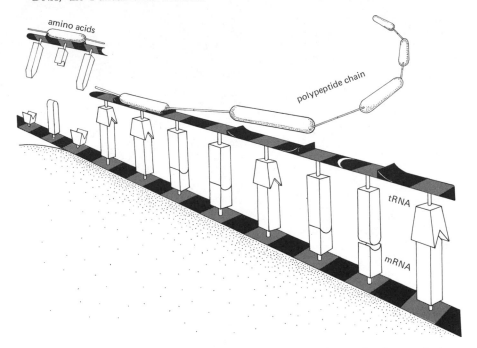

amino acids

polypeptide chain

tRNA

mRNA

4.11 The amino acids brought by the *t*RNA create polypeptide chains which themselves form proteins. Once the *t*RNA has served its purpose, it breaks away and repeats the process. Note that it is the DNA in the nucleus which dictated the subsequent events.

triplets that do not code for particular amino acids, but act to interrupt the transcription process. Exactly how this entire process is controlled so that cells function the way they are supposed to is only beginning to be understood; it is very complicated and we need not go into it here. But it is important to see that the change in even a single base pair in the DNA of a sperm or egg cell could change the nature of a particular code for an important protein, and cause serious, perhaps deadly, problems for the individual concerned. In its simplest form, that is what a mutation is.

Mutations. In the copying process, wherein the gene reproduces itself, three kinds of change can occur (Figure 4.12). First, one of the three bases in the triplet can be lost. This is called a deletion. Second, a different base can be substituted. Substitution results in coding for a different amino acid, or coding for none at all, depending on the particular substitution. Finally, a new base can slip in—this is called an insertion or addition. Any combination of deletion, substitution, and insertion can occur. As bases are read three at a time, insertion or deletion can shift the reading frame, so that the new combination will call in different amino acids.

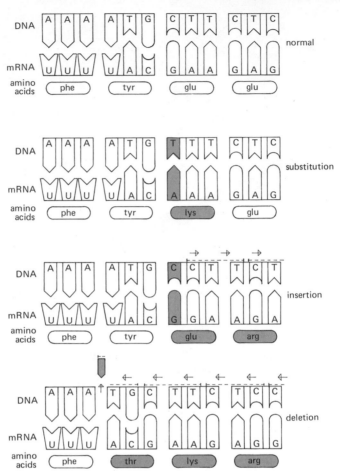

4.12 Three types of missense point mutation. (Redrawn from Lerner, I. Michael. 1968. *Heredity, evolution and society.* San Francisco: Freeman, p. 86.)

These changes within the gene are called point mutations because they occur at specific points—namely, in the bases. Point mutations are the most important sources of the genetic variation that is significant for evolution, at least in animals. The supply of new genetic material represents necessary insurance for the future of the species. If the environment should change, a species that had been perfectly adapted to the old environment, but with an inflexible genetic makeup, might die out.

All genes have a spontaneous mutation rate, which differs from one gene to another and which is itself under genetic control. The rates of spontaneous mutation for single genes are very low, between 1 and 5 per 100,000. (Another estimate gives 1 per 500,000 for all genes.) Most mutations are harmful, and only a small percentage of them are beneficial. The spontaneous mutation rate for a given gene represents the best compromise that evolution has reached in balancing the need for new genetic material against the harmful effects (ranging from slight handicaps to death) of most mutations. [Figure 4.13]

The spontaneous mutation rate is controlled by specific genes, called mutator genes, as well as by some mechanism, as yet unknown, within each cell. In addition, there are many external agents which influence the mutation rate—mostly increasing it, although a few are said to decrease it. Those agents which increase mutations are heat, radiation, and certain

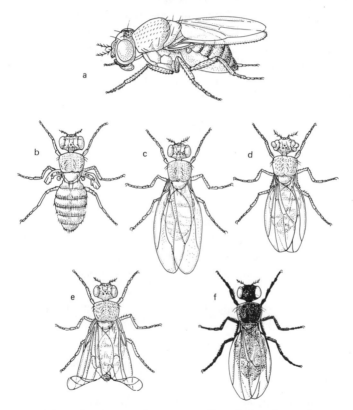

4.13 *Drosophila melanogaster*, the common fruit fly has been of great value in the study of genetics. Possessing only four pairs of chromosomes and producing new generations every ten to fifteen days, scientists have been able to study changes in eye color, wing size, and other genetic variables which come about through normal breeding. Spontaneous mutant forms as well as artificial mutations induced by X rays (above) may also be studied. Illustrated are (a) *Drosophila* (the normal form), (b) wingless, (c) long-winged, (d) bar-eyed, (e) curly-winged, (f) ebony-bodied.

chemicals. Because of its importance, radiation will be dealt with at length in Chapter 16. Here I might point out that the chemical mutagens, as they are called, do not seem to be a great threat to man as yet, although this may change. Caffeine, for example, which most of us take in the form of coffee and tea, and theobromine, the main active constituent of cocoa, resemble the purine bases, adenine and guanine, in chemical structure. Caffeine and theobromine are mutagenic to bacteria and the fruit fly, *Drosophila*, but large amounts fed to mice cause no increase in point mutations.

Concerning the effects of heat on the mutation rate, it is known that wearing trousers increases scrotal temperature by about 3° Celsius (C), which could almost double the mutation rate. But before you rush to exchange blue jeans for kilts, you might reflect that hot baths, which the Japanese take every night, or the sauna or steam bath, to which the Finns are addicted, raise scrotal temperature even more than trousers do, and there is no evidence that the Japanese or the Finns have higher mutation rates than people who never bathe.

Then what, really, is a *gene?* Most molecular biologists would respond that a length of DNA that codes for a complete chain of amino acids (a polypeptide chain) best suits that term. We can speak, then, of the gene that codes for the alpha hemoglobin chain, an important constituent of the red blood cells.

The genes are linearly arranged in enormously long chains of DNA. If the DNA strands of a single human sperm or egg cell were all stretched out, end to end, they would be only 2-millionths of a millimeter thick, but they would stretch almost four feet from beginning to end. Of course, they are not so arranged, but are in separate chromosomes. The entire blueprint for a particular human being is contained in the DNA of the fertilized egg, and is reproduced in almost all subsequent cells which constitute the individual.

SUGGESTED READINGS

Bracket, J. 1961. The living cell. *Scientific American*. September.

Levine, Robert P. 1968. *Genetics*. 2d ed. New York: Holt, Rinehart and Winston.

Mendel, G. 1948. *Experiments in plant hybridisation*. Cambridge, Mass.: Harvard University Press (translation of Gregor Mendel's original paper, first published in 1866).

Merrell, David, J. 1975. *An introduction to genetics*. New York: Norton.

Wallace, Bruce. 1966. *Chromosomes, giant molecules and evolution*. New York: Norton.

5
THE HUMAN COMPLEMENT OF CHROMOSOMES

Chromosomes are linear bodies within the cell nucleus which become visible during certain stages of cell division (see Figure 4.2). The word *chromosome* means "colored body" and refers to the fact that these threadlike bodies take up certain stains which make them visible under the microscope. Since cells normally contain a great deal of water, which is colorless, special stains are needed to disclose details of cell structure. Genes are physically strung out along the chromosomes like beads along a necklace. This arrangement is best seen in the giant chromosomes found in the cells of salivary glands among some insects, notably *Drosophila*, the fruit fly. Distinctive bands and puffs along the length of the chromosome correspond to genes known to be present from examination of the insects and from breeding experiments.

Basics of Chromosomes

Each species has a characteristic number of chromosomes, which do not lose their identities. Chromosomes are invisible during the resting phase of the cell because they are relatively uncoiled and stretched out thin. During cell division, they become concentrated into more tightly wound coils which are thicker and therefore easier to see. The number of chromosomes in a human body cell is forty-six (Figure 5.1); the chimpanzee

a   Chromosomes of a human male

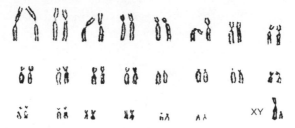

b   Chromosomes of a human female

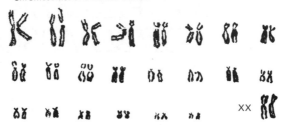

c   Chromosomes of a gorilla male

5.1 Representative karyotypes of the human male and female and the male gorilla. The great majority of animal species appear to have a characteristic number of chromosome pairs. In man this number is twenty-three; in the gorilla, twenty-four. The sex chromosomes appear at the extreme right of the third line of each karyotype. The chromosomes appear to correspond closely in detail between man and gorilla, much more so than between man (or gorilla) and representative monkeys. (Redrawn from Klinger, H. P., Hamerton, D. M., and Lang, E. M. The chromosomes of the hominoidea, in Washburn, Sherwood L., ed. 1963. *Classification and human evolution*. Chicago: Aldine, p. 237.)

has forty-eight; the rhesus monkey, forty-two. Other primate species have chromosome numbers that vary between thirty-four and seventy-two. The fruit fly has only four, which makes it much easier to study than man.

The chromosome number in man consists of forty-four autosomes and two sex chromosomes, called the X and Y chromosomes—either two Xs or an X and a Y. The autosomes have to do with all parts of the body; the sex chromosomes determine sex as well as carrying other genes which also affect various parts of the body. A person with two X chromosomes is a female; an X and a Y is a male (Figure 5.1). The chromosomes are paired; one member of each pair comes from the father and one from the mother. In the same way, a single set of twenty-three chromosomes is passed on to the offspring. But the set passed on is not exactly like that received from either father or mother. In the production of germ cells, which are called sperm in a male, eggs or ova in a female, the members of each pair of chromosomes are distributed at random. This means that the germ cell which an individual produces and which carries the hereditary contribution that individual makes to his offspring, will carry some chromosomes originally derived from the individual's father, and some derived from the individual's mother. Furthermore, during the production of germ cells, a process called meiosis, the two members of a pair of chromosomes may exchange material by the mechanism of crossing over (Figure 5.2). This means that even a single chromosome which you pass on to your offspring may carry genes from both your mother and your father. The sexual process is thus constantly reshuffling or recombining genes, and the possible number of combinations from two parents is enormous. This is why every single individual in the world, now alive or who has ever lived, with the exception of identical twins, is absolutely unique. There never has been anyone like you or me, and there never will be.

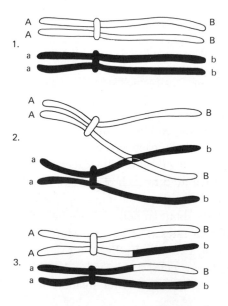

5.2 Crossing over: 1. Toward the end of prophase I of meiosis (Figure 5.5a), two chromosomes, each with two chromatids, appear together.

2. Two adjoining, nonpaired chromatids cross over to form a chiasma or point of exchange at corresponding loci.

3. The chromosomes separate following the exchange of portions of two chromatids.

Crossing over increases the number of different genetic combinations that any given mating can produce, and therefore contributes greatly to the variability in a population.

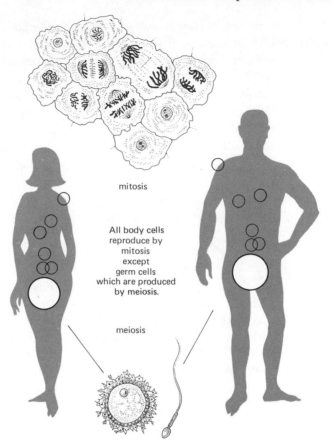

5.3 Mitosis and meiosis. With the exception of sperm and egg cells which are produced by meiosis, all body cells are produced by mitosis. The schematic drawing between the two figures shows tissue cells in various steps of nuclear (mitotic) division as they might appear on a microscope slide.

Reproduction: Mitosis and Meiosis. Chromosomes are passed along in two ways, during mitosis and meiosis, depending on the kinds of cells in which they occur (Figure 5.3). In body cells, each of the forty-six chromosomes splits down the middle, reproducing itself exactly (except for occasional errors). One of each pair is distributed to each of the two daughter cells produced, which in this way keep the same number, forty-six chromosomes. This process, mitosis, lasts about an hour (Figure 5.4). Different cells in the body divide at different rates. The life of a white blood cell is about two weeks; a red blood cell, two months; liver and skin cells also divide fairly often. Connective tissue cells divide only about fifty times during a person's lifetime of seventy years. After infancy, nerve cells never divide and a small fraction dies every day. To the extent that the number of brain cells increases intelligence, we can personally only lose a little basic intellect each day. The catch is, of course, that total numbers are less important than functional ability.

Sperm begin to be produced in adolescence and are produced at a fairly constant rate thereafter, in astronomical numbers, for a man's life-time. A single ejaculation contains 200 to 300 million sperm. It takes about two months for a sperm to develop from the cell which gives rise to it. The egg, on the other hand, has a very different life history (Figure 5.6). All of the cells which will develop into eggs are formed in the female fetus by the age of five or six months of intrauterine life, and they even begin meiosis before birth. Their total number is about 750,000, and they must last for a woman's entire reproductive lifetime. Most eggs degenerate at various times in a woman's life, during the course of their maturation. Until the beginning of reproductive life, starting between twelve and thirteen years of age on the average in this country, these egg-forming cells have the diploid chromosome number, forty-six. After the woman reaches sexual maturity, each month, under the influence of pituitary hormones, a few cells complete meiosis, halving their chromosome number to the haploid number, twenty-three. One or two cells may become fertilized. In that case they complete meiosis, redistributing the haploid, or halved, number of chromosomes before the two nuclei, of sperm and egg, actually fuse. This monthly maturation of a few ova ceases at the menopause, around age fifty.

Thus, the full course of sperm production takes two months, but the full course of mature egg production takes twelve to fifty years. For this reason most chromosomal abnormalities appear to originate in the mother rather than in the father. Much more can go wrong over the longer period of time in which the chromosomes are in a resting state. And, as we shall see, the longer the duration of this resting state—that is, the older the mother—the more frequent are such abnormalities.

Chromosomal Mutation. Genetically, chromosomes are the physical link between the generations. The physical mechanisms of chromosome replications, which occur in mitosis, and of random assortment and of independent segregation, which occur in meiosis, reshuffle the genes in each generation. As already noted, this recombination of genes already present accounts for a much greater fraction of the variation in any species, at a given time, than does gene mutation, a rare event. Nevertheless, the only source of new variation is mutation. There is another class of mutations besides the point mutations of genes: those of chromosomes. By a variety of mechanisms, chromosomes or parts of chromosomes can be

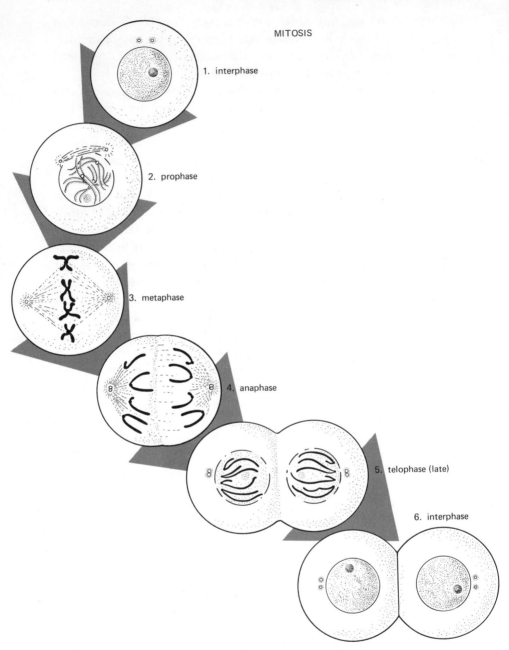

MITOSIS

1. interphase

2. prophase

3. metaphase

4. anaphase

5. telophase (late)

6. interphase

5.4 Mitosis is the process of cell division in which chromosomes are duplicated and distributed equally to two daughter cells so that each of the latter has genes and chromosomes identical in kind and number to the original parent cell. The following phases are arbitrary divisions in what is actually a continuous process.

added, lost, or changed in structure. I have already referred to crossing over (Figure 5.2) in which corresponding portions of replicated chromosomes become exchanged. This is almost as common as the transmission of intact chromosomes, with all genes linked in their original orders.

Chromosome Anomalies and Medicine

The frequency of chromosomal anomalies is considerable: one in five hundred births has an aberrant number of sex chromosomes, and one in four hundred has an aberrant number of autosomes. The combined total for all abnormalities of chromosome number is about 0.5 percent of all births, roughly twenty-five thousand persons per year in the United States. In addition, a large percentage of spontaneous abortions and miscarriages can be attributed to abnormalities of the larger autosomes which are incompatible with life, and many have abnormalities of the other chromosomes as well.

Techniques of Chromosome Counting. Chromosome counting and identification rests on five technical advances: (1) Colchicine, a chemical which stops cell division, is applied in a stage favorable for counting. (2) A hypotonic, or low-salt, solution is used, to make the chromosomes swell and the cell membranes burst. (3) Preparations are squashed between two

---

1. Interphase: The nucleus is clearly visible; the chromosomes cannot be seen. At least one nucleolus is visible. Centrioles are here represented above the nucleus.

2. Prophase: The chromosomes gradually appear and thicken by coiling themselves. They are of two strands held together by a centromere. Toward the end of this phase, the chromosomes begin to move toward the center of the cell. The centrioles begin to move apart. The nucleus membrane begins to disappear.

3. Metaphase: With the centromeres at the equator, the chromosomes align themselves down the middle of the cell on spindle fibrils which the centrioles, now at opposite sides of the cell, have spun between them. (In these diagrams, there are only two pairs.)

4. Anaphase: The centromere divides and the centrioles pull to their respective sides the separated sister chromatids.

5. Telophase: The chromosomes uncoil and lengthen. The nuclear membrane forms. Nucleoli appear and the centrioles replicate. Meiosis is now complete. The two new cells, like their parent cell, are diploid; that is, they possess two of each type of chromosome. The cells return to interphase when the chromosomes again become less distinct.

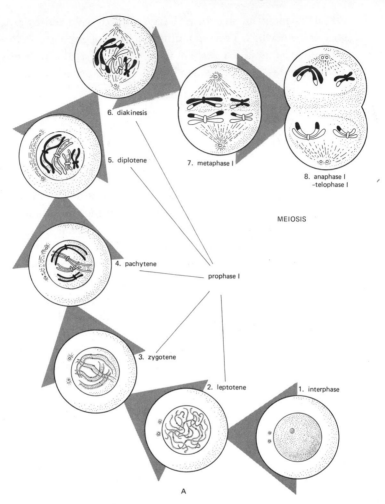

6. diakinesis

5. diplotene

7. metaphase I

8. anaphase I
   –telophase I

MEIOSIS

4. pachytene

prophase I

3. zygotene

2. leptotene

1. interphase

A

5.5 Meiosis is the reduction division in the production of gametes (egg or sperm) during which the diploid chromosome number is halved. Two nuclear divisions occur while the chromosomes themselves divide only once. The process is divided into the following phases:

1. Interphase: The nucleolus is visible but the chromosomes are indistinct.
2. Leptotene: The first stage of prophase I. The chromosomes appear as fine separate strands.
3. Zygotene: The chromosomes begin to pair (a major difference between meiosis and mitosis). This pairing which occurs down the whole length of the chromosome is not fully understood.
4. Pachytene: Pairing is now complete. The centrioles begin to make their spindle fibrils (the chromosomes become more and more tightly coiled throughout the whole of prophase I).
5. Diplotene: The chromatids of the paired homologous chromosomes are now

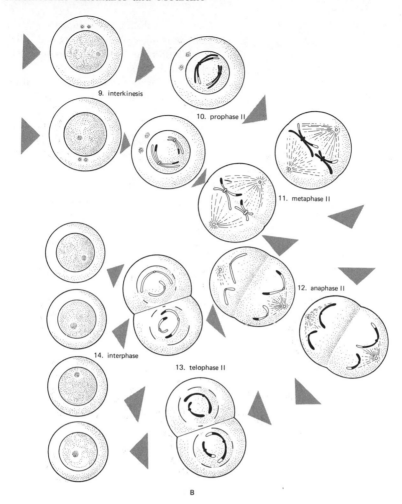

9. interkinesis

10. prophase II

11. metaphase II

12. anaphase II

14. interphase

13. telophase II

B

clearly visible although they are not separated from each other. At this stage crossing over may begin to take place.

6. Diakinesis. The chromosomes are now shortened and the nuclear membrane begins to disappear.

7. Metaphase I: Pairs of chromosomes move to the equator of the cell. The centromeres do not divide.

8. Anaphase I–Telophase I: Double-stranded chromosomes move apart to opposite centrioles. The nuclear membrane appears and the two cells divide.

9. Interkinesis: No reproduction of genetic material is apparent. Note that these two cells are haploid, that is, they have only one of each type of chromosome.

10. Prophase II–14 Interphase. This second phase is essentially mitotic except there is no pairing since the nucleus is haploid and there are no homologous chromosomes.

The four new cells are haploid containing single-stranded chromosomes.

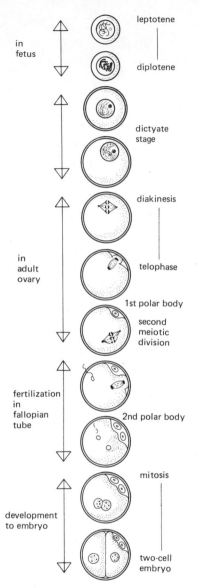

leptotene

in
fetus

diplotene

dictyate
stage

diakinesis

in
adult
ovary

telophase

1st polar body

second
meiotic
division

fertilization
in
fallopian
tube

2nd polar body

mitosis

development
to embryo

two-cell
embryo

5.6 The volution of a human egg: The first four steps of the first meiotic division (leptotene through diplotene) take place in the ovary of the fetus.

Meiosis is arrested late in the diplotene stage (before the birth of the fetus) when a nucleus called the germinal vesicle forms in the egg. The egg itself becomes enclosed in a follicle where it remains throughout childhood until after puberty. Then it is stimulated by hormones to resume meiosis.

Progressing through diakinesis, metaphase, anaphase, and telophase, the first meiotic division ends with a large daughter cell and a smaller polar body. The second meiotic division begins as soon as the first ends. In humans, as in most mammalian species, ovulation occurs when the egg is in metaphase II.

If the egg has been fertilized, the second meiotic division is completed in the fallopian tube.

glass surfaces, to spread the chromosomes apart. (4) Cells are cultured in artificial media to permit continuous observation of single cells and sampling of their descendants. Photographs are taken, cut up, and the chromosome pairs are matched by eye. (5) Finally, a number of new stains have been developed very recently, making it possible to identify individual chromosomes by characteristic patterns of banding. This supplements the earlier methods based on chromosome size; position of the centromere, which unites the strands of each member of a pair; and the presence or absence of small satellites.

Barr Bodies and Sex. In 1949, M. L. Barr and Bertram, two Canadians, made an accidental discovery. They were studying the effects of electrical stimulation on the nerve cells of the cat. In examining the altered cells under the microscope, they noted a distinctive mass of material, staining like chromosomes, inside the nucleus of most cells of some cats, but in none of the cells of other cats. It turned out that only female cats had this nuclear mass, renamed sex chromatin—and now called the Barr body (Figure 5.7). These observations have been extended to other mammals, including man, and to other tissues of the body, including skin and blood. It is now possible to determine with complete reliability the nuclear, or genetic, sex of normal persons. This is usually done from scrapings of the inside of the cheek, hair roots, or blood cells. Thirty to sixty percent of female cells have the Barr body, which is invariably absent in normal males. The Barr body follows the n-1 rule: there is one less Barr body than the number of X chromosomes in the cell. The Barr body is the inactivated X chromosome postulated by Mary Lyon in her famous hypothesis, which we shall discuss presently.

How can this information be used? Its most important application is in human intersexuality, where patients may have both male and female gonads as well as discrepancies between external and internal genitalia, or between genetic and anatomical sex. In some such cases, sex diagnosis by nuclear chromatin can be lifesaving, where early diagnosis can lead to treatment with adrenal hormones. Other uses of nuclear sex determination are to diagnose sex before birth (by aspiration of amniotic fluid), to calculate the sex ratio of aborted fetuses, to study the fate of tissue grafts, to trace the origin of certain tumors of the sex cells, and to distinguish fetal from maternal components of the placenta.

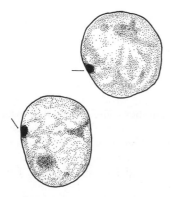

5.7 Nuclei of skin cells from a normal female, showing X-chromatin bodies (Barr bodies). (Redrawn from Moody, P. A. 1975. *Genetics of man.* New York: Norton. Note, Moody drew his from photomicrographs in Barr, M. L. 1960. Sexual demorphism in interphase nuclei. *American Journal of Human Genetics* 12:118–127.)

Nondisjunction and Errors of Number. In 1959, a French team headed by J. Lejeune electrified the medical world by showing that in mongolism, the commonest single type of severe mental defect (more correctly known as Down's syndrome), there are forty-seven instead of forty-six chromosomes (Figure 5.8), the extra one being a small autosome with a satellite, chromosome number 21. This condition of three instead of two of chromosome number 21 is called a trisomy. People who have it show a characteristic appearance, with short stature and a number of facial features superficially resembling those of an Oriental. Several other trisomies have been identified, each with its own constellation of physical features. Most are severely deficient mentally and die soon after birth.

Attention turned next to two conditions showing a discrepancy between the external genitals and nuclear sex, as seen in the Barr body count. It turned out that Klinefelter's syndrome had the chromosome constitution *XXY*—that is, a male with an extra *X* chromosome. These patients are tall, have breast development like females, small testes, and do not produce live sperm. Persons with Turner's syndrome have the *XO* constitution—a single *X* chromosome and no *Y*. They are incomplete females, with short stature, webbed neck, and undeveloped uterus and ovaries.

The mechanism causing all these anomalies is nondisjunction, which is the failure of the chromosomes to separate on cell division. During the stage of meiosis in which the paired chromosomes separate from the midplane and proceed to each end of the spindle (the anaphase), one

5.8 The chromosome set of a male with Down's syndrome. Called trisomy 21 due to the presence of an extra chromosome 21 (in box). The abnormality, which occurs in approximately 1 of 650 births, is thought to be caused by nondisjunction, the failure of homologous chromosomes to separate from each other following synapsis in meiosis. Likewise, daughter chromosomes may fail to separate from each other in the anaphase of mitosis. (After Bearn, Alexander G. and German, James L.)

chromosome may lag behind and so appear twice in one of the daughter cells and not at all in the other. Depending on the chromosome content of the other germ cell which the abnormal cell meets at fertilization, the chromosomes of the new individual may be incompatible with life (no X chromosome) or may produce one of the various above anomalies; a Turner's syndrome (XO), Klinefelter's syndrome (XXY), or other less common sex-chromosome anomalies. A missing X chromosome or autosome is incompatible with life. [Figure 5.9.]

Methods of Discovery in Chromosome Anomalies. Four different scientific disciplines contributed clues to the puzzle of mongolism, or Down's syndrome. Mongoloids had been known since their first scientific description by Langdon Down, a British physician, in the 1860s. They were abundant in institutions for the mentally retarded, but there was no good explanation for their condition. Medical clinicians had also recognized two kinds of persons who had developed abnormally in their sexual development, an incomplete male (Klinefelter) and an incomplete female (Turner), both sterile. Quite independently, three other groups of scientists were working along lines that ultimately converged. Geneticists working with the fruit fly, *Drosophila*, noted that nondisjunction occurred occasionally and led to intersexes, or sexually abnormal, sterile flies. The frequency of nondisjunction was greatly increased by radiation and with increasing maternal age.

Epidemiology is the study of the distribution and determinants of disease in man. Epidemiologists noted three things about mongolism that suggested a genetic component: (1) the marked increase of mongolism with maternal age, (2) its aggregation in some families, and (3) its concordance in identical twins. They further noted that leukemia was about ten times more common among mongolians than in the general population. Also, radiation increases the incidence of leukemia. So here were several clues pointing to some common factor: maternal age and radiation increased nondisjunction in *Drosophila;* maternal age and leukemia were associated with mongolism; radiation increased leukemia.

			number of X chromosomes			
		0	1	2	3	4	5
number of Y chromosomes	0	not found	Turner	normal female	normal or sterile	female	female
	1	not found	normal male	Klinefelter	Klinefelter	Klinefelter	Klinefelter
	2	not found	normal	Klinefelter	Klinefelter	Klinefelter	Klinefelter

5.9. Summary of the kinds of abnormalities involving sex chromosomes in man.

This made it logical to look for nondisjunction in the chromosomes of mongolians.

But it is one thing to study nondisjunction of the four chromosomes in *Drosophila,* and quite another in man, whose chromosome number was then not even known, much less individual chromosomes identified. This had to wait for the cytologists to develop the technical advances discussed earlier. The first breakthrough was the discovery of the Barr body, then the accurate counting and identification of chromosomes. It was soon found that both Turner and Klinefelter syndromes showed a discrepancy between nuclear and genital sex. The Turner female should have had a Barr body, but had none; the Klinefelter male should have had none, but had one. So the search was on for chromosome disturbances to explain all three conditions—mongolism and the two syndromes, Klinefelter's and Turner's. Within a few months of each other, in 1959, the solutions were found.

Further developments led to identification of other chromosomal aberrations of both the sex chromosomes and the autosomal trisomies. Two investigators in Philadelphia, Nowell and Hungerford, decided to look more closely at the chromosomes in leukemia patients. They looked first, logically enough, at acute leukemia, the kind that is associated with mongolism, and found nothing. They might have quit here, reported their negative finding, and received an A for a well-conceived and well-executed piece of research. But on a hunch, they went on what is sometimes called with great scorn by methodological purists, a fishing expedition. They examined chromosomes in chronic (myeloid) leukemia, which is rarely if ever found among mongolians. They got a D for study design, but their fishing expedition landed a whale—the first distinctive marker of a human cancer cell. They found, in the circulating blood cells, but not in other cells, and in some patients but not all, a small chromosome, which they named the Philadelphia chromosome. It is a deleted chromosome number 22—that is, a chromosome number 22 minus its long arm. It is not known where the missing fragment went, whether it was lost or became translocated onto another chromosome. Newer staining methods should tell us soon. Since the discovery of the Philadelphia chromosome, another distinctive chromosome was found in a small fraction of patients with chronic lymphocytic leukemia. This is called the Christchurch chromosome, from the New Zealand city where it was discovered, and represents deletion of the short arm of a group C chromosome (one of a large group of medium-sized chromosomes of similar appearance, one arm being somewhat longer than the other).

A further example of scientific method again comes from epidemiology. Not content to stop with the usual association between mongolism and maternal age, two British epidemiologists, Carter and Evans, computed the chance of a mother's having a second mongol child after having had one already. To their surprise, the risk was greatest in the youngest mothers, under age twenty-five—just the opposite for the risk of mongolism generally. This trail led back to the chromosomes. Under the microscope another mechanism for mongolism was discovered—namely, the translocation of one chromosome number 21 onto a large autosome, usually number 15 resulting in one number 21 and one large autosome which represents the other 21 and a 15 joined together. The mothers are phenotypically normal, since they have the correct total amount of chromatin; but at meiosis, the translocation chromosome number 15-21 may appear in a germ cell together with the normal chromosome number 21. When fertilized by a normal sperm with one chromosome number 21, the resulting offspring will have three such chromosomes and will be a mongolian. So, whereas most mongolians, those born to older mothers, are chance events, a few, born to younger mothers, have a chromosomal basis which is predictable from examination of the mother's cells. The practical value of this discovery is enormous. The abnormal chromosome is easily identified. A young mother who has had one mongolian child can be counseled as to the probability of producing another mongolian. If her chromosomes are normal, she can be reassured that the chance is extremely small. If her chromosomes are abnormal, the chances are between 10 percent (observed) and 50 percent (theoretically expected) that her next living child will be a mongolian. These odds may be more than she is willing to accept.

The X and Y Chromosomes in Action. A great variety of other sex chromosome anomalies has been identified, some with as many as five X chromosomes, with or without a Y. The most interesting, as well as the most numerous, after the Klinefelter and Turner types, are the triple X, or superfemale, who is apparently normal and fertile, although somewhat retarded mentally, and the XYY, or supermale, who is tall, has acne, and has been suspected of a tendency to be violently antisocial, although there are reservations as to the solidity of this association—at any rate, there are plenty of normally behaved XYYs.

The study of persons with abnormal sex chromosomes has greatly increased our knowledge of the normal function of the X and Y chromosomes. The X chromosome, as we saw, is necessary for life; no humans have been found who lack it; that is, the hypothetical YO individual has not been found. The X chromosome also carries a great deal of genetic information. It has at least seventy-five known specific mutated genes, which with its general endocrine and other feminizing traits are well known.

The Xg blood group system, so called because it occurs on the X chromosome, was first found in Grand Rapids, Michigan. It is the first blood group definitely linked to any chromosome. The antigen Xg^a behaves like an X linked dominant. This discovery is a great importance for genetic linkage studies and for mapping the X chromosome.

The Y chromosome is far from inert, however. Regardless of how many X chromosomes are present, persons with a Y chromosome always have testes. Without a Y chromosome, the phenotype is always female—in the XO (Turner) syndrome and in the three, four, and five X patterns as well as the normal XX. The only clear single gene now known on the Y chromosome is called the HY gene. It produces a protein which tends to disrupt tissue transplants from males to females, and may also influence the differentiation of the male genital organs. It was discovered some years ago in mice, and has only lately been demonstrated in human males. A possible Y-borne mutant gene is one alleged to produce hairy ears in some men. Factors on the Y chromosome retard skeletal maturation, which is a male characteristic.

The Lyon Hypothesis. Formulated by several workers, including Mary Lyon, at the same time in 1961, this hypothesis states that among females, part or all of one X chromosome in every somatic cell is genetically inactive, or relatively so, and goes to form the sex chromatin or Barr body. The decision as to which X chromosome, paternal or maternal, will be the inactive one in a particular cell is made in the first weeks after conception, since the Barr body can be identified at about the sixteenth day. It is random which X chromosome in a given cell is to be the inactive one. Once the decision is made in a given cell, all descendants of that cell abide by the decision and have the same X chromosome inactive. The germ cell line of the normal female does not participate in this process of fixed differentiation of the X chromosome—only somatic cells do. (Germ cells are the line preserved from the original zygote which gives rise to sperm and egg cells, while somatic cells are those which differentiate to form all the tissues of the body.)

In the normal female, the Lyon hypothesis provides a mechanism for dosage compensation. This means that the female, with two X chromosomes, has no more of certain proteins, produced by genes on the X chromosome, than males with only one X. Dosage compensation also occurs in the anomalous conditions with multiple X chromosomes and multiple $(n-1)$ Barr bodies, such as the three, four, and five X chromosome states. Due to dosage compensation, resulting from inactivation of all but one X chromosome, the consequences of multiple X chromosomes are not nearly as severe as for the other trisomies (numbers 13, 17, and 21) which involve much smaller chromosomes. Dosage compensation is not present in the Y chromosome, however, Men with the rare XYY constitution produce more of the Y-borne and HY proteins than do normal XY men.

The Problem of Artificial Mutagens. A final topic concerns the effects of mutagens on the human chromosomes. Each generation holds in trust the genetic heritage of the species. As a manner of speaking, nature's concern is more for the species than for the individual. The living are important, in an evolutionary sense, only to the extent that they preserve and transmit the genes that they hold on a temporary basis—on loan, if you like. One of our great concerns today is whether any aspects of our environment, which is increasingly man-made, damage this precious genetic heritage.

We have already seen that some naturally occurring agents, such as heat, ultraviolet light, and ionizing radiation, increase the mutation rate of genes. The same agents increase the rate of chromosomal mutations by breakage and nondisjunction. Human beings have greatly increased the amount of ionizing radiation to which we are all exposed, in such forms as medical X rays, luminescent watch dials that glow in the dark, and nuclear energy. We have also added some synthetic mutagenic chemicals and increased our exposure to those that occur naturally. This is a complex matter. Mutations are random. Some are beneficial, but most are harmful. The real problem is whether the potential and problematic future benefit of mutations for the species outweighs the certain harm to individuals and to the species, here and now represented by the defective persons produced by mutations.

On the whole, and recognizing our present ignorance, our verdict would have to be that increased mutation rates are harmful to the human species and should be avoided. Nature has developed protection against naturally occurring mutagens. The famous rats who live on the radioactive sands in Kersala, India, and who receive eight times as much radiation as neighboring rats, show no apparent genetic damage. But we cannot afford to wait for nature to develop protection against man-made mutagens. The first commandment of clinical medicine holds true in this field as well: *"Primum non nocere"*—The first thing is to do no harm.

SUGGESTED READINGS

DuPraw, E. J. 1970. *DNA and chromosomes.* New York: Holt, Rinehart and Winston.

Lerner, I. Michael, and Libby, William J. 1976. *Hereditary, evolution, and society*. 2d ed. San Francisco: Freeman.

McKusick, V. A. 1971. The mapping of human chromosomes. *Scientific American*. April.

Mazia, D. 1961. How cells divide. *Scientific American*. September.

Moody, Paul Amos. 1975. *Genetics of man*. 2d ed. New York: Norton.

6
GENES IN INDIVIDUALS AND POPULATIONS

Genes in Individuals and Families

Dominance and Recessivity. In 1866, Mendel wrote, "Characters which are transmitted entire, or almost unchanged in hybridization, and therefore in themselves constitute the character of the hybrid, are termed the dominant, and those which become latent in the process, recessive." If we substitute the word *heterozygote* for *hybrid,* this is still the definition today. Ideally, dominant traits are those expressed in the heterozygote, as well as in the homozygote; recessive traits are expressed only when homozygous, being masked in the heterozygote. Many rare traits in man, as well as some common ones, are distributed in families according to Mendel's definition. The specific pedigree pattern depends on whether the genes for these traits are dominant or recessive, and whether they are carried on an autosome or on a sex chromosome. Unfortunately, except for anomalies, visible features in man which act like the characters of Mendel's famous peas, or even like some features of coloration in other animals, are not known. The example of eye color, frequently used, is not as straightforward as has been thought, and I shall use it on an as if basis, and accept the assertion that brown eyes are dominant over blue. Then we have two alleles (genes of different effect at the same chromomosome locus) on an autosome, which may pair thus (*B* for brown is dominant; *b* for blue is recessive):

BB	Bb	bb
brown	brown	blue

By Mendel's laws, the following matings are possible:

$BB \times BB$ offspring brown only (BB = 0% blue)
$BB \times Bb$ offspring brown only (BB, Bb possible = 0% blue)
$BB \times bb$ offspring brown only (all Bb = 0% blue)
$Bb \times Bb$ (equal numbers of BB, Bb, bB, bb = 25% blue)
$Bb \times bb$ (equal numbers of Bb, bb = 50% blue)
$bb \times bb$ (offspring blue only, all bb = 100% blue)

Dominance and recessivity are artificial concepts, depending on the accuracy of our methods. The better our methods, the more likely we are to distinguish the homozygote from the heterozygote. In sickle-cell anemia, for example, the anemia is recessive. A person needs the double dose to become anemic. The heterozygote, who has the gene in single dose, has the sickling trait. He is not anemic, but his red blood cells will sickle (change from round to crescent or other abnormal shapes) under certain conditions, notably lack of oxygen, as at high altitudes or when exposed on a glass slide.

The blood groups show another kind of dominance, called codominance, in which both characters are expressed in the heterozygote. Persons with blood type AB show this. It is now clear why we must distinguish between genotype and phenotype. In dealing with a dominant and a recessive trait, we cannot tell from the manifestation, or phenotype, whether a person is homozygous or heterozygous. In the case of brown eyes, we cannot tell whether the genotype is BB or Bb. In the case of blood group A, which is dominant to O, we do not know whether a person with group A is AA, the homozygote, or AO, the heterozygote. We can tell the genotype only by analyzing pedigrees.

Sex Linkage. Dominance and recessivity hold for genes located on the X chromosome as well as for those on the twenty-two pairs of autosomes. But their expression is different in men and women, because men have only one X chromosome, whereas women have two. The Y chromosome does determine male sex, the slower masculine rate of skeletal maturation, and the HY protein described earlier, but its other functions remain to be discovered. Any gene on the X chromosome—and about seventy-five or more separate ones have now been identified—will be expressed in the male, whether it is dominant or recessive (Figure 6.1). In the case of a recessive trait, like hemophilia or red-green color

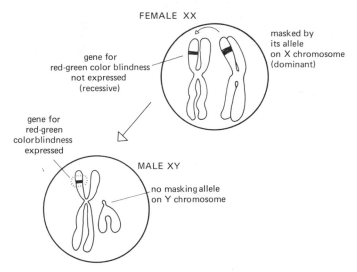

FEMALE XX

gene for
red-green color blindness
not expressed
(recessive)

masked by
its allele
on X chromosome
(dominant)

gene for
red-green
colorblindness
expressed

MALE XY

no masking allele
on Y chromosome

6.1 Sex-linked inheritance in man. The effect of a recessive gene (here for color blindness) on a female X chromosome is usually masked by a dominant normal gene on its corresponding allele on the other X chromosome. (It is rare that the same recessive gene will appear on both X chromosomes.) Should this recessive gene be passed to a son, however, it will be expressed since there is no masking allele on the Y chromosome.

blindness, there will be no normal allele on the paired X chromosome to mask its effect. (An X-linked dominant will of course be expressed in either sex, the same as a dominant on an autosomal chromosome.) A woman who has an X chromosome with a gene for hemophilia will show that trait only if her other X chromosome has the same recessive gene—a very rare event. Usually, the recessive gene is masked by its normal allele on her other X chromosome. That is why such women transmit the gene for hemophilia to half their sons and half their daughters, but neither they themselves nor their daughters are bleeders (Figure 6.2).

Genetics of
Normal Variation

Human variation, we have noted, like the variation in all species, has two components, genetic and environmental. So far, we have been concerned with the genetic component, but increasingly we shall be dealing with environmental factors. Strictly speaking, characteristics are neither inherited nor acquired. What is inherited is the capacity of a given genetic makeup

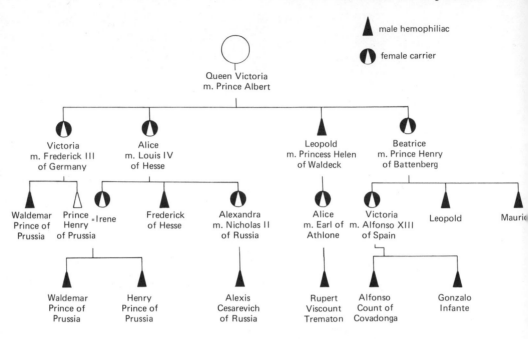

male hemophiliac

female carrier

Queen Victoria
m. Prince Albert

Victoria
m. Frederick III
of Germany

Alice
m. Louis IV
of Hesse

Leopold
m. Princess Helen
of Waldeck

Beatrice
m. Prince Henry
of Battenberg

Waldemar
Prince of
Prussia

Prince
Henry
of Prussia
= Irene

Frederick
of Hesse

Alexandra
m. Nicholas II
of Russia

Alice
m. Earl of
Athlone

Victoria
m. Alfonso XIII
of Spain

Leopold

Mauric

Waldemar
Prince of
Prussia

Henry
Prince of
Prussia

Alexis
Cesarevich
of Russia

Rupert
Viscount
Trematon

Alfonso
Count of
Covadonga

Gonzalo
Infante

6.2 The afflicted descendants (only) of Queen Victoria. Hemophilia, a sex-linked recessive disease, was passed to the members of many of Europe's ruling families by Queen Victoria. (Photo: courtesy Warder Collection.)

to react or respond in certain determined ways to normal and abnormal environmental conditions. The limits to possible variation caused by changed environmental factors are set by the genetic makeup. For example, starvation or malnutrition will stunt the growth of children and result in small adult size. Feeding will increase the height of the next generation, because the genotype is not changed, only the phenotype. Beyond a certain point, set by the genes, no amount of environmental improvement will increase height. No matter how much food you stuff into a pygmy child from birth on, he will never grow as tall as a Watusi or even to medium size. For the same reason, within recent Harvard families of fathers—born between 1910 and 1919—and their sons, the sons are no taller than their fathers (Figure 6.3), whereas the height of the general American population continues to increase. Harvard families reached an optimum environment, as far as height is concerned, one generation ago, whereas the general population has not yet reached it. Even when such an optimum environment is reached, genetic variation in height will re-

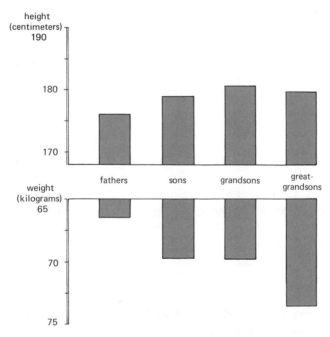

6.3 Trends in height and weight through four generations of families at Harvard from a study made by the author. The fathers on the average were born in 1858 and successive generations in 1888, 1918, and 1841. Height shows a leveling off in the second generation. It is not known whether the continued increase in weight represents fat or muscle. (SOURCE: Damon, A. 1968. *American Journal of Physical Anthropology*, n.s. 29:45–50.)

main. This illustrates a very important principle, which at first seems paradoxical: the more uniform the environment, the larger is the fraction of total variation due to heredity. Contrariwise, the more uniform the heredity, the more can we attribute variation to the environment.

Polygenes. Genetic variation is of two main kinds, discontinuous and continuous. Hemoglobin is normal or sickle-type (see Figures 1.16 and 1.17), blood group is *O, A, B,* or *AB.* This is discontinuous, single-gene variation. On the other hand, height, intelligence, skin color, and blood pressure vary in small steps, continuously. These traits are measured, whereas the single-gene traits are counted. Continuous variables are determined by many genes, with cumulative or additive effects; continuous variables are polygenic or multifactorial. Furthermore, the polygenic traits are strongly influenced by the environment, in great contrast to the single-gene traits. However you tinker with the environment, you will not change a person's eye color or blood group. But height, intelligence, and blood pressure are all greatly affected by the environment.

This next idea is very significant in the study of human biology: as we see them now, polygenic traits are of more immediate significance to mankind than single-gene traits. Our criterion of importance here is quantitative, not ethical or personal. Without question, a harmful mutation, a harmful recessive trait, or a trisomy is tragic to an individual or a family. But the sum total of all the death, disability, and economic cost due to single-gene disease is but a tiny fraction of that due to the diseases of multiple causation. These are the polygenic diseases with a strong environmental component. Heart disease, strokes, cancer, diabetes, and arthritis are just a few examples. And even in the area of normal function, quite apart from disease, such important characteristics as size, shape, strength, speed, skill, and various mental abilities are all polygenic.

The methods for studying polygenes are largely statistical. Some of these methods were developed earlier, and others later, than the pedigree analyses of classical single-gene genetics, begun by Mendel. Polygenic distributions follow the bell-shaped normal curve. They are studied by comparing the averages, and the spread around the average, as in the example for height (Figure 3.3). One might also see whether identical twins are more alike in height than fraternal twins of the same sex, thereby obtaining a rough indication of the relative importance of heredity and environment in determining height.

As an illustration of the nature of polygenes, let us consider the complexity of the interaction between genotype and environment for the polygenic trait, skin color. Looking around the world, we find differences in skin color among populations, among families, and among individuals.

The major races of mankind differ in skin color. Between and within populations of European origin there is considerable variation (and the same is true of darker peoples also). Some people are pale, others are swarthy, with all degrees of gradation in between. There is a genetic component to this variation. In unexposed areas of the skin, Europeans originating farther away from the equator are generally lighter than those whose origins are closer to the equator. That is, Spaniards and Italians are darker than Norwegians and Swedes. Within families, children generally resemble their parents in skin color, and children of unlike parents are generally intermediate. Within a population, males are generally darker than females, in unexposed areas of the skin. All of these differences—between individuals, between the sexes, among families, and among populations—have a genetic basis.

The presence of a continuous series of intergrades strongly suggests that skin color is not controlled by a simple set of only a few genetic factors. There is, to be sure, a recessive gene which, in homozygous form, produces an albino—that is, a person with no pigment whatever. But albinism is very rare, about one per twenty thousand births, and it has nothing to do with the range of pigmentation found in the general population. Rather, it is a mutation blocking the formation of normal pigment altogether (see below).

So far, we have mentioned only genetic influence on skin color. But superimposed on the genetic influence is a strong environmental effect. Except for albinos, nearly all persons, however light or dark their skins, darken on exposure to sunlight—that is, they tan. The effect of tanning is of course most noticeable among light-skinned persons. This means that among Europeans, phenotypic skin color is not always a reliable guide to the genotype.

Now let us take a closer look at the factors influencing skin color within the individual, to get some idea as to their number and variety. Human skin has two major layers, the dermis or inner layer, and the epidermis, or outer layer (Figure 6.4). In turn, the epidermis has an outer layer of protective cells, which is constantly being shed and replaced by new cells pushing up from below. In the deeper layer of the epidermis are cells called melanocytes containing a dark brown pigment, called melanin, in the form of distinct granules called melanosomes. Each pigment cell manufactures these granules and injects them into a number of adjacent epidermal cells, a process which is constantly going on. Every stage of this process is under genetic control. In addition, the whole process is also subject to environmental factors, notably sunlight, as we have seen. Without reviewing the detailed biochemical steps leading to the

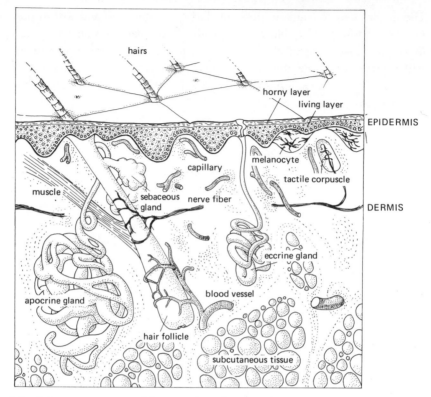

6.4 Schematic section of human skin showing the epidermis (horny layer of dead surface cells plus underlying layer of living cells), the dermis (containing blood vessels, nerves, and glands) and fat-rich subcutaneous tissue. Melanocytes (shown here in surface view as through teased away from the epidermis) are evenly distributed deep in the epidermis. They are irregularly shaped cells which produce the dark pigment melanin. Granules of melanin (called melanosomes) are injected by the melanocytes into the surrounding epidermal cells which they protect by absorbing the ultraviolet rays of the sun. Skin color is determined not by the number of melanocytes (which is roughly the same for all human beings regardless of race) but by the amount of pigment which is manufactured.

Apocrine glands are scent glands which secrete in response to emotional stress. Eccrine glands are sweat glands of which humans have between two and five million. (From various sources.)

synthesis of melanin we may simply recognize that they are complex. The starting point is an amino acid, tyrosine. An enzyme, tyrosinase, converts tyrosine to a substance whose long name, dihydroxyphenylalanine, is shortened to DOPA. DOPA becomes an even more complex substance called a quinone. Many quinones join together, or polymerize, to form melanin. This process, which is controlled by additional enzymes, occurs within the pigment granules, or melanosomes, so that the final composition of the granules is a combination of melanin and protein.

That every step of this process is under genetic control has been demonstrated by finding persons who, whether by mutation or by inheritance, lack one of the factors required for the complete process. People who lack the enzyme tyrosinase are albinos; they have melanosomes and tyrosine, but cannot convert tyrosine to melanin. Albinism is a well-known autosomal recessive trait, as we have seen. Other people, with all the right tyrosine and tyrosinase, have melanosomes that function poorly, on a genetic basis. They develop white patches on their skin, called vitiligo. Not only is melanosome function controlled genetically, but so also is the shape of the melanosome, which is round in redheads and oval in everyone else. The number and size of pigment granules are also under genetic control. Melanosomes are more numerous and larger in darker persons, families, and populations than in lighter ones. Even the packaging of melanosomes within the cells is under genetic control. Among Australian aborigines and Africans, each pigment granule is large and single, surrounded by its own membrane; among Caucasians, Orientals, and American Indians, the melanosomes are smaller, and they come in multiple packets of two, three, or more, surrounded by a single, common membrane. In addition to all this, the activity of melanocytes in injecting neighboring cells is under genetic control. Of all these structural and functional influences which I have been listing, some are all-or-none, like the presence or absence of tyrosinase in albinism, but most are graded in extent.

This might be a good place to emphasize a view of gene action and mutation. The gene giving rise to albinism is not a gene *for* albinism; rather, like many other mutations, it is a deficiency gene, an allele of the normal gene which produces the enzyme tyrosinase. Two normal genes are typically present, but one gene is enough to produce the enzyme and so masks the presence of the deficiency gene in the heterozygote. Thus the latter is recessive, and albinism appears only when it is homozygous.

Other factors in the individual which affect skin color, and which are genetic, include the concentration of pigment cells in various parts of the body. In unexposed skin, the darkest regions are the genitals and the areolar regions around the nipples of the breasts, in both sexes. Another genetic variable is the thickness of keratin, the outer, transparent layer of the epidermis. The keratin transmits variable amounts of sunlight, and thereby stimulates the pigment cell to a variable extent. Hormones have a strong influence on skin color; in fact, the pituitary gland produces a specific skin-darkening hormone called melanocyte-stimulating hormone. Another hormone, progesterone, produced by the ovary, also darkens the skin, so that a small percentage of pregnant women or those taking contraceptive pills show darkening of the skin. Still another gland, the adrenal, is involved in normal pigmentation. Hypofunction of the adrenal, called Addison's disease, produces a bronzing of the skin and pigmented patches in the mucous membranes. For all three glands, pituitary, ovary, and adrenal, the amounts of hormones produced, and the sensitivity of the melanocytes to those hormones, are both under genetic control.

All this may be telling you more about skin color than you really care to know. But the point is to show that there must be very many genes interacting to determine the level of pigmentation in any one individual. It is impossible to say exactly how many enzymes, structural proteins, and control substances take part in the system; a reasonable guess would be two to three dozen. To picture a polygenic system that controls pigmentation, we should not think of a simple-minded model of thirty or forty genes, each contributing one unit of darkening. Differences in the effectiveness of the many interacting enzymes, structural proteins, and control substances can be combined in different ways to produce a graded series of genotypes and corresponding phenotypes. Therefore, estimates that as few as five pairs of additive genes control skin color are doubtless too simple.

Genes in Populations

Hardy-Weinberg Law. The next step is to think less of genes in pairs, as in individuals, and more of the total gene picture in a population—the gene pool, as it is called. The genotype of an individual is his particular pair of genes: of the *A*, *B*, and *O* blood genes, he can have only two. The gene frequencies of a population are the proportions of all the alleles, or genes of a series, in the whole gene pool, from which the genotypes of individuals are drawn. For a population, the frequencies, as proportions, might read, $A = 0.3$, $B = 0.2$, $O = 0.5$.

The cornerstone of population genetics is the Hardy-Weinberg law. Like all scientific laws, this one makes assumptions; we will ignore them here in discussing the law; in the next chapter we recognize their vast importance. The Hardy-Weinberg law states that without disturbing influences such as mutation or selection, in any breeding population the proportions of genes, and genotypes, remain constant. This is the

Hardy-Weinberg equilibrium. The existing gene frequencies determine the frequencies of the genotypes: knowing either, one can compute the other, as is illustrated below.

The mathematical statement is simple enough. Let us pursue the case of two genes, or alleles, only. The proportions of genes (and also of genotypes) sum to 1.00: thus $p + q = 1$. Suppose our population has 80 percent of genes B for brown eyes, and 20 percent of genes b for blue eyes (remembering that this is grossly oversimplified, since eye color is more complex). Then $p = 0.8$, and $q = 0.2$, adding to 1.0. From this, the proportion of genotypes must be as follows:

Ova

$B = p\ = 0.8$ $b = q = 0.2$

Sperm

$B = p = 0.8$ $BB = p^2\ = 0.64$ $Bb = pq = 0.16$

$b = q = 0.2$ $bB = pq = 0.16$ $bb = q^2\ = 0.04$

Adding these gives:

$p^2 = BB = 0.64;$ $2pq = Bb = 0.32;$ $q^2 = bb = 0.04$

Or, for *phenotypes:*

brown $= 96\%$ blue $= 4\%$

Now it can be seen that $(p^2 + 2pq + q^2)$, as obtained above, is by simple algebra the expansion of the binomial: $(p + q)^2$, so that merely by substituting the gene frequencies, 0.8 and 0.2, one can figure the proportions of the genotypes at once, without making a table as above.

This is the simplest way of looking at the Hardy-Weinberg law. However, we usually do not know gene frequencies in a population to begin with and, in fact, computing gene frequencies from genotypes, not the reverse, is the main object of the Hardy-Weinberg principle. We simply do the above calculations backward. In the above case, we would note that the frequency of the recessive homozygote (blue eyes, which we know to be bb in genotype—brown eyes might be BB or Bb) is 4 percent, or 0.04. This represents q^2 in the expression above. The square root, q, is therefore 0.2, the frequency of gene b; so the frequency of gene B is $p = (1 - q)$, or 0.8. QED.

These calculations are not restricted to two allelic genes only. Take the ABO example suggested above, and set $p = 0.3$, $q = 0.2$, and $r = 0.5$. Then by $(p + q + r)^2$, the genotypes are:

p^2	q^2	r^2	$2pq$	$2pr$	$2qr$
AA	BB	OO	AB	AO	BO
0.09	0.04	0.25	0.12	0.30	0.20

And, combining into phenotypes—for example, $AA + AO$—the percentages of the blood groups are:

O	A	B	AB
25%	39%	24%	12%

Such is the meaning of recessive genes that although O is the most frequent gene, blood group A is the commonest phenotype. The reverse calculations, from phenotype to gene frequencies, are considerably more complex than in the case of two genes only.

Applications of the Hardy-Weinberg Principle. Why should we want to do these calculations, except for idle curiosity? The Hardy-Weinberg law has two broad areas of application, which we might call evolutionary and genetic. Its evolutionary implication, granting the conditions under which it applies, is that the genetic picture shown by a breeding population many generations ago was much the same as it shows today. In fact, it takes only one generation of random mating to establish such an equilibrium, even if two previously distinct breeding populations should mix. So, assuming that a group continues to interbreed at random and does not mix with its neighbors and that the genes involved do not affect reproduction or the survival of children, the frequency of genes will remain constant over the generations. This is roughly the case for the ABO blood groups. Blood groups therefore provide a useful tool for tracing the recent history of human migrations and the relationships among peoples. For example, Melanesians in the Southwest Pacific have the dark skins, frizzly hair, broad noses, and other physical features of Africans. In fact, these Pacific islanders were at first called Oceanic Negroids. But blood groups, among other characteristics, are so different that the two populations cannot have had any contact in recent evolutionary times.

The other area of application of the Hardy-Weinberg law is genetic. First of all, the law settles two questions which arose early in the history of genetics. Why do not dominant genes drive out recessive ones? Or, why doesn't the mixture of the two produce a blend, a single intermediate type? The Hardy-Weinberg law states that the genes remain separate, as Mendel had found, and that they maintain their original proportions in a population, unless the equilibrium is disturbed.

As to calculating gene frequencies from those of genotypes, we wish to do this for four reasons:

1. Gene frequencies are better than phenotype frequencies for comparing a trait in different populations. For example, it is better to know the gene frequencies for blood types A and B than the frequency of persons with A and B blood, since the A phenotype can have either the AA or the AO genotype, and the B phenotype can be either the BB or the BO genotype.

2. Gene frequencies allow us to test the mode of inheritance of a given trait.

3. They allow us to calculate the expected genotype frequencies of children from any type of marriage.

4. They enable us to study gene frequencies over time to see whether they remain constant. If so, the system is in equilibrium; if not, we look for the sources of instability.

But all these genetic reasons for interest in the Hardy-Weinberg law pale before the final one: It allows us to calculate the proportion of heterozygotes when the frequency of the recessive phenotype is known.

In the real world, we do not know the gene frequencies in a population. What we see is people—people with distinctive physical or biochemical or behavioral characteristics. In other words, we see phenotypes. We count their frequency in the population. Observation of these distinctive persons and of their families tells us whether these phenotypic traits have a genetic basis; and if so, what kind. That is, a certain trait will cluster in families and possibly in males or females. Further analysis may show that the trait is an X-linked dominant or recessive. Or pedigree analysis may show that the trait follows an autosomal recessive type of inheritance.

Once we know that a trait is genetic and determined by a single gene, we can determine the genotype frequency from the phenotype frequency. To take a simple example, an autosomal recessive, the phenotype is the genotype. All persons who show the trait must be homozygous recessives, with a double dose of the gene, one from each parent. In our example, if four persons in one hundred of the population have blue eyes, the frequency of the blue-eyed phenotype is 4 percent; this is the same as the frequency of the blue-eyed genotype. Now we have two further problems: one is to find the proportion of brown-eyed persons who are homozygous dominants, with two genes for brown eyes, one from each parent. The other problem, by far the most important, is to find the proportion of heterozygote carriers—that is, brown-eyed persons who carry, in recessive or hidden form, the recessive gene for blue eyes. The reason we are so particularly interested in heterozygotes is that they are the main source of recessives in the new generation. With random mating, there are very few matings between two persons with a rare recessive trait, even assuming that those with the trait survive to reproductive age. In our example, even when as many as one gene in five, or 20 percent of the genes, are for blue eyes—only 4 percent of the population will have blue eyes—and this is a much higher percentage than one finds for rare recessives like albinism or PKU, each of which occurs in about one per ten thousand births in populations of northern European origin. Even cystic fibrosis, the commonest condition caused by a single-gene autosomal recessive, occurs in only one in thirty-seven hundred births. Because rare recessives are usually harmful, and because the largest source of such recessives are the heterozygote carriers, we want to be able to estimate their numbers and to identify heterozygote carriers for purposes of genetic counseling, biochemical testing, and possible prevention of the disease.

It is here that the calculations using the Hardy-Weinberg law, already shown on page 97 became so useful. Suppose that, in the example, the genes are not those for brown and blue eyes, but for the normal and the sickling genes for hemoglobin, A and S. This is something very real for the black population of the United States, although the suggested frequencies of $A = 0.8$ and $S = 0.2$ would be more likely in an African population. As in the example, the recessive, the sufferers from sickle-cell anemia, would be 4 percent. How frequent are the carriers, the heterozygotes? From the same figures, of the 96 percent "normals" in that population, 64 percent would be AA, with no sickling gene, and 32 percent would be carriers, those who are protected against malaria but who, if married to one another, have a 25 percent risk of bearing an anemic child.

Now, at last, we must recognize by name the assumptions underlying the Hardy-Weinberg law. They are these five: (1) random mating, (2) no mutation, (3) no selection, (4) no drift, and (5) no gene flow. No human population comes anywhere near this idealized picture. In the real world, feathers and bullets do not fall at equal speeds, but the law of gravity is very useful. Likewise the Hardy-Weinberg law, although far from describing the situation for any real human population, is useful and sometimes comes surprisingly close to the actual gene frequencies.

SUGGESTED READINGS

Dunn, Leslie, C. 1959. *Heredity and evolution in human populations.* Cambridge, Mass.: Harvard University Press.

Friedlaender, J. S. 1975. *Patterns of human variation.* Cambridge, Mass.: Harvard University Press.

Lewontin, R. C. 1974. *The genetic basis of evolutionary change.* New York: Columbia University Press.

Montagna, William. 1965. The skin. *Scientific American.* 212: 56ff.

Stern, Curt. 1973. *Principles of human genetics.* 3d ed. San Francisco: Freeman.

Wallace, Bruce. 1968. *Topics in population genetics.* New York: Norton.

7
EVOLUTION AND SELECTION IN MAN

Assumptions of the Hardy-Weinberg Equilibrium

The preceding chapter ended with a listing of the five assumptions of the Hardy-Weinberg principle—five perturbing factors that would not allow the equilibrium to persist. They might all be rolled into a single assumption: no evolution. For evolution, major or minor, is no more than a change in the status quo, a modification of gene frequencies, and thus a movement in the physical nature of a population or a species. In this chapter we shall review these assumptions, now become agents of evolution, giving heavy emphasis to selection.

Mutation and the Fate of Genes. As we have seen, every gene mutates at a fairly constant rate, spontaneously. The average for any gene is about one per one hundred thousand gametes per generation. Probably every one of us carries at least one or two highly harmful recessive genes in the heterozygous state. Our average genetic load is three to eight so-called lethal equivalents—that is, on the average we carry many genes, each with slightly harmful effects. Their combined effect would be equivalent to three to eight recessive genes that in the homozygous state would result in death before reproduction. (Much more frequent, as we have seen, are the chromosome mutations, with about one per thousand births being trisomic—that is, having three instead of two chromosomes in some pair. We ignore them in discussing the Hardy-Weinberg law, which deals with genes.)

Mutation is going on all the time. The chance that a mutation will affect the particular chromosome locus under consideration in Hardy-Weinberg calculations is quite small. For lethal recessives, that is, those which cause death in the homozygous form, at equilibrium the mutation rate is exactly the same as the proportion of genotypes. That is, if the frequency of a rare lethal recessive is one in ten thousand, this must be the mutation rate—otherwise, the removal in each generation of homozygous recessives which result from the mating of heterozygote carriers would change the gene frequencies in each generation, and hence the genotype frequencies. We would not then have an equilibrium.

Some lethal or near-lethal conditions are so common that their persistence over time would require impossibly high mutation rates. Take cystic fibrosis, the commonest lethal single-gene disease of childhood. About one in every three thousand white births has cystic fibrosis (it is very rare among blacks). To maintain equilibrium between genes added by new mutation and those lost by early death of homozygotes, before reproducing, would require a mutation rate of one in fifteen hundred gametes, or germ cells. This is incredibly large, in view of what we know about mutation rates. What are other possibilities? There may be several loci rather than one, with a mutation at any one of these loci resulting in the phenotype of cystic fibrosis. But even if you could imagine ten such loci, you would still have a mutation rate of one in fifteen thousand—still much too large. The more likely explanation is heterozygote advantage; that is, the heterozygote is reproductively favored over the homozygous normal. Such reproductive success could come about in two ways: increased fertility and improved survival. Both of these have been shown actually to obtain for heterozygous carriers of another rare lethal recessive gene, Tay-Sachs disease.

Heterozygote advantage, you will recall, is just what we find with sickle-cell anemia; the heterozygote, who has both normal and sickle hemoglobin, has an advantage over the homozygous normal. Heterozygote advantage is one of the mechanisms of balanced polymorphism (Figure 7.1). *Balanced* means that the system is in equilibrium. In balanced polymorphism, the supply of recessives is maintained in higher frequency than mutation alone would provide. Balanced polymorphism is the situation where more than one allele persists over time in a population. The human blood groups are an example of a polymorphic locus, sometimes balanced and sometimes not. There must be some advantage to having several *ABO* and Rh blood group variants, an advantage strong enough to outweigh the very real disadvantage of blood type incompatibility between mother and fetus, which causes an appreciable percentage of fetal and newborn deaths.

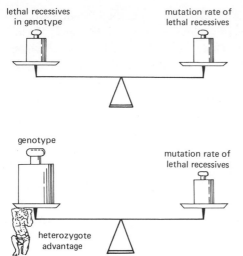

7.1 Balanced polymorphism. Heterozygote advantage refers to the benefits derived from having, in single form, what would be deleterious or lethal in a double (homozygous) dosage. In the drawing, the heterozygote advantage supports the greater number of these recessives found in the genotype than could be explained by the mutation rate. Without heterozygote advantage, the proportion in the genotype must equal the rate with which they are produced by mutation.

It is estimated that 30 percent of all human loci are polymorphic. The orthodox view is that all of them represent mutations that have proved advantageous in some past environment, and some may prove useful in future environments. A newer view is that some may always have been selectively neutral.

Heterozygote advantage refers to single genes. The corresponding term for polygenes, or for the organism as a whole, is *heterosis*. When two inbred stocks of plants or animals are crossed, the offspring tend to be larger, healthier, and more fertile than either inbred parent strain. One reason is that harmful recessives are masked by the normal alleles at many loci. Evidence for heterosis in man is equivocal. There are some reports that offspring of parents, of the same population but from different villages, are taller than those whose parents were born in the same village. But direct measures exhibit no such effect. Infants born in Hawaii to parents from different major racial stocks have shown no effect on birth weight or mortality.

Gene Flow. This mechanism also subverts the Hardy-Weinberg principle. It reflects the fact that the human race is a single, interbreeding species. Local populations can remain in partial or total isolation, genetically speaking, for many generations. But no group of *Homo sapiens* has remained isolated for periods long enough to produce a new species; there has always been enough gene flow, from migration of individuals or groups, to prevent speciation. A new species, by definition, could not naturally breed with present-day man and have fertile offspring.

Genetic Drift and the Founder Effect. Drift is a third antagonist of the Hardy-Weinberg equilibrium, occurring precisely because mankind does tend to break up into breeding isolates. Geography, religion, and culture all produce breeding isolates in which chance differences from one generation to another can affect gene frequencies markedly (Figure 7.2). The Hardy-Weinberg law states a mathematical expectancy, not a certainty. If you flip a coin ten times, you expect five heads and five tails as the commonest single outcome, or the average result; but some runs will be very different, say two against eight; and so it is with genes. In such a change, however, genes differ from coins; a coin keeps its fifty-fifty chance of heads or tails, but a gene shift from five-and-five to two-and-eight becomes the new equilibrium, so that drift back toward five-and-five is less likely than a final zero-and-ten, or the total loss of one gene. The smaller the group, the stronger the likelihood of drift. For example, most American Indians have virtually 100 percent blood group *O*. But

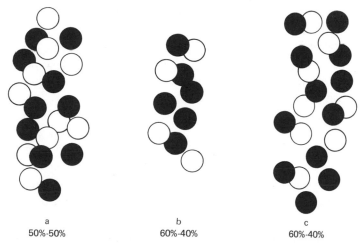

a
50%-50%

b
60%-40%

c
60%-40%

7.2 Various kinds of selection or genetic drift change the proportion of genes. Suppose in the parent generation (a) two genes are present in approximately equal proportions. Suppose, too, that in the next generation only a small portion of these are in the parental gene pool (b) for the succeeding generation (c). (This is an important principle of Darwinism. It has been estimated that approximately 50 percent of conceptions do not pass their genes on to another generation [Figure 7.5].) Suppose also, however, that selection has favored whatever it is the dark gene causes very slightly, that is, just enough to cause a six-to-four ratio rather than a five-to-five ratio. (Note that this change could also have been brought about by chance fluctuation or drift. In so small a population, a relatively slight accident could produce this excess of two dark genes.) While the increase in the parental generation is slight, the next generation (c) will have a clear majority of dark genes (twelve-to-eight). Unless light genes become very heavily selected for in succeeding generations, it is unlikely that the earlier fifty-fifty ratio will be re-established.

certain tribes have a high frequency of A. The Blackfoot and Blood Indians have up to 80 percent of blood group A, as compared with 2 percent in the Ute Indians. This is probably due to drift.

As mentioned earlier, the founder effect is a special case of drift. It is particularly striking in small, isolated populations. So up to this point we have seen that three of the assumptions of the Hardy-Weinberg law are violated in real human populations. These three—mutation, gene flow, and drift—concern the constancy of gene frequency over time.

Types of Mating in Man

The major assumption, no selection, also concerns gene frequency, and I will come to it shortly. Let us first consider random mating, or panmixia, which is still another assumption. In theory, at a given locus, any gene would meet any other of its allelic group by the laws of chance alone. But in the real world, this is nonsense. Human populations have never bred randomly. All people have incest taboos on the mating of close relatives— father-daughter, mother-son, brother-sister, and the like. In addition, there are three major deviations from random mating: inbreeding, selective mating, and assortative mating.

Inbreeding. This involves the restriction of mating to a given gene pool (Figure 7.3). This is breaking down in modern life, but much of the world still continues its ancient patterns of breeding only or mainly within a given area, linguistic group, religious group, or ethnic group. Many surviving tribal groups actually prescribe marriage between cousins. Inbreeding increases homozygotes at the expense of heterozygotes, for all genes in a population.

Selective Mating. This occurs when some characteristic is prized in both sexes, and persons possessing it are therefore at a reproductive advantage. Health has always been desirable; so have physical vigor, strength, skill, a smooth skin, and harmonious physical development. Intelligence is now thought to be such a characteristic in the modern world. This could explain the several-point increase in IQ shown among Scottish children between 1932 and 1947. To take another example, skin color: in Japan and India light skin color is prized in a mate of either sex. Among the isolated Wabag of the New Guinea Highlands, dark-skinned persons are preferred. To the extent that such traits have a genetic basis, they will increase in populations practicing selective mating. That is, not only phenotypes, but also gene and genotype frequencies will change.

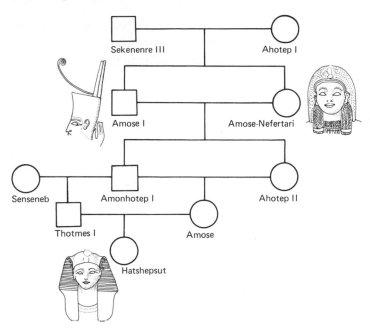

7.3 The Pharaohs of Egypt. Marriage between brother and sister not only was acceptable but was favored among the dynastic families of Egypt. Hatshepsut, the great Queen of 1500 B.C. was the product of at least four generations of brother-sister mating corresponding to the strict inbreeding of genetic experimenters. If we consider her mother, Amose, who was descended from three generations of brother-sister matings (without the half-brother-sister parentage of Hatshepsut) it can be seen that the ancestor loss is extreme: two great-grandparents instead of the possible eight, and a corresponding loss of chromosomal variety.

One kind of selective mating is sexual selection, in which certain traits are prized in one or the other sex. Among ourselves, tallness in men seems to be such a characteristic, much more so than in other societies. In a series of several thousand Harvard men, followed for many years after college, those who married were indeed taller than those that never married.

How large an effect has selective mating had, or will it have, on human evolution? Except for intelligence, it is unlikely that selective mating will have more than a minor, locally variable effect. The ideals sought in mates vary too much from time to time and from group to group.

Homogamy. The third major deviation from random mating is assortative mating, or homogamy, in which like marries like. Strictly speaking, this is positive assortative mating—one would have negative assortative mating if people avoided mates like themselves. Despite the folk saying that opposites attract, there are no demonstrated cases of negative assortative mating in man, although redheads are alleged to marry one another less often than would occur by purely random mating.

Assortative mating for age is widespread in primitive as well as advanced societies. Obviously this would not hold for groups where old men customarily married much younger wives, but such groups are the exception. In addition, Europeans and Americans mate assortatively for intelligence, for height and its associated body dimensions, and for hair and eye color. None of the few other populations studied in this way mate assortatively for body size—not the Japanese, not American Indians, not Solomon Islanders. On the other hand, there is assortative mating for skin color in the American Indians and in the Solomon Islanders.

The explanation for assortative mating remains obscure. Arctic gulls of different subvarieties have distinctively colored rings around their eyes. They mate only with gulls whose rings are the same color as their own. The basis for this is early imprinting on the mother gull. Of course, man has invented the mirror, so he knows what looks like himself.

Whatever the underlying psychodynamics, which must remain speculative, the net result of assortative mating is like that of inbreeding. Without changing total gene frequencies in a population, both inbreeding and assortative mating increase homozygotes at the expense of heterozygotes. Homozygotes, at the extremes of the range of polygenic traits, are an important source of variability in a population—and populations need genetic variability to permit a range of adaptations to changed environments. As we shall see, the predominant trend in modern life is to reduce genetic variation and the scope within which natural selection can act. Any mechanism that increases variability is therefore useful to the species.

Selection in Man

We have now covered four of the five agents whose absence is assumed by the Hardy-Weinberg law, and we find that all occur constantly in human populations, contrary to the assumptions. So does the fifth, selection, which acts through several mechanisms and leads to several kinds of outcome, all of which influence the frequency of genes and genotypes in the population.

What is selected? The cardinal feature of selection is that it works on phenotypes. We talk about gene frequencies, but we cannot see them, and neither can nature. We see people. Nature sees people. People do the adapting, the mating, the living, and the dying. Genes for superior intelligence or skill will be lost if they appear, for example, in a child who also has two sickling genes. He dies young, and with him go his superior genes.

Natural Selection. Of the several types or mechanisms of selection in man, natural selection has gone on for the longest time. As we have seen, fitness in the biological sense is defined in terms of the contribution made to the genes of the succeeding generation. Selection means differential fitness. It works via differences in reproduction or in survival to reproductive age—in other words, via differential birth rates and death rates. Factors in the environment influencing selection include disease, food supply, and physical features of the habitat, such as climate, altitude, and light. The biochemical and physical adaptations to these environmental factors will be dealt with more completely later in the book. But briefly, there are some long-term trends that have no apparent purpose, such as loss of third molar teeth, loss of hair on the middle phalanx of our fingers, and increasing roundness of the head. Other long-term trends have a real purpose, such as increase in brain size. But in general, geneticists now believe that most genetic characteristics serve some function, or did at one time, even though we may not know what that function was. As W. F. Bodmer puts it, they are "relics" of previous selective crises.

The *ABO* blood groups are a good example. We really have no idea why there are four blood types, *O, A, B,* and *AB* in most human populations. As mentioned, they cause fetal death in a small number of cases, by incompatibility between mother and fetus, so they must have some compensating advantage. They cause trouble in transfusion, but this was certainly no factor in their origin and persistence. They are associated with a group of diseases of the upper gastrointestinal tract—patients with peptic ulcers have higher frequencies of blood group *O,* those with stomach cancer have more blood group *A.* But these are diseases that occur late in life, often after reproduction has been completed, so they are not likely to have played much of a role in natural selection.

Figure 3.4 shows a diagram of the distribution of blood groups *A* and *B* in 215 human populations. If blood groups occurred at random, one would expect an even distribution throughout the area of the graph. But instead we see a concentration between 15 and 30 percent for *A* and between 5 and 20 percent for *B,* so that *O* can vary from 50 to 80 percent. Evidently some combinations of alleles must have greater adaptive value than others. The spread indicates further that natural selection does not merely rank these alleles on a single scale of adaptive excellence, rejecting the two of inferior rank. Here again, we see that the human species is inherently polymorphic and variable. We do not know what function the blood groups serve—and this is true also of the Rh and many other systems. But there is no doubt that natural selection has worked on them. It has been suggested that certain blood groups confer resistance to smallpox and plague, those ancient scourges of man, but the evidence is quite shaky.

Artificial Selection. What other kinds of selection are there? We have already mentioned sexual selection. Artificial selection is that imposed by man, either directly, as in prescribed breeding patterns, or indirectly, in modifying the environments to which men are selected and adapted. Selective breeding would include prescribed marriages of royalty or of members of Indian castes, who have hereditary skills which may rest on a genetic substrate. Wars, conquests, and slavery alter breeding patterns markedly. The ancient Chinese held competitive examinations in which poor but brilliant boys could enter the Mandarin class, marry the daughters of mandarins, and pass on their genes. Societies which provide social mobility via their educational, religious, or military systems do the same. On the other hand, the several religions which impose priestly celibacy have the opposite effect, losing a portion of their presumed genes for character and educability in every generation.

Changes in the man-made environment since the invention of agriculture and the domestication of animals have imposed new conditions for selection. Crowding increased the spread of infectious microorganisms and parasites, to which populations have had to adapt. There are marked population differences in susceptibility to disease as a result of previous exposure—not only during an individual's life, but also in the population's past history. The most striking examples are the worldwide decimation of previously unexposed populations by tuberculosis, smallpox, and even measles. Modern cities, with crowding, overstimulation, and disruption of normal biological cycles, are a new environment, something that will be discussed later in the book.

Other man-made changes in gene frequency result from medical advances, permitting the survival and reproduction of those who would otherwise succumb to genetic disease (Figure 7.4), such as diabetes, cystic fibrosis, phenylketonuria, or sickle-cell disease. This sweeps them under the rug, as it were, bequeathing to future generations a large genetic load dependent on a stable social order. Social policy, in the form of taxation, welfare, and birth control, can have a profound effect on birth rates, just as medical and public health measures affect the death rate. To the extent that the persons affected differ genetically, there will be change in gene frequencies.

Effect of the Human Environment. We should also recognize the reduced scope for natural selection brought about by lowered birth and death rates in advanced countries. We have now reached the point where

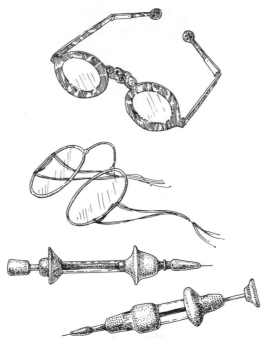

7.4 Medical advances permit the survival and reproduction of those who would otherwise succumb to genetic disease. Early eyeglasses and native syringes from China. (Eyeglasses: Redrawn from a photograph, courtesy the American Museum of Natural History, New York. Syringes: Redrawn from a photograph, courtesy Pitt Rivers Museum, Oxford, England.)

most people survive and reproduce small families. So both birth selection and death selection must work within a much smaller range of variation than before. On the other hand, this reduced variation affects only about half of all fertilized eggs, or zygotes, those that reproduce. It is estimated that close to half of all zygotes never reproduce (Figure 7.5). About 15 percent are lost before birth, 3 percent are stillborn, 2 percent die shortly after birth, 3 percent die before maturity, 20 percent never marry, and 10 percent of those who marry remain childless. Thus there is still considerable room for selection. Genotypic differences that favor survival to reproduction at any of these stages will enjoy an advantage.

There are various types of selection occurring in modern man. Directional selection occurs when one gene, one combination, or one end of the normal curve of distribution for polygenes has selective value, that is, is favored in survival and reproduction. When an environment changes, a previous adaptation to the previous environment may become maladaptive. The species will then evolve in such a direction as to adapt to the

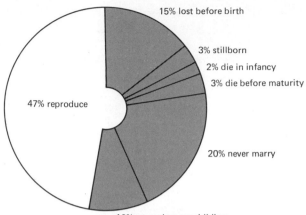

7.5 Over half of all individuals conceived do not reproduce.

new environment. Examples of directional selection are long-term changes like those toward roundness of head, increased head size, thinning of skull, and possibly—if the age at menarche continues to fall—earlier maturation.

Stabilizing, or normalizing, selection maintains a currently successful adaptation. This is the force which preserves the present successful and characteristic phenotype of a species, since by definition most individuals fall around the mid-region of the normal curve of distribution. One kind of selection, for the individual, is longevity. We have been following a cohort of twenty-five hundred men who were measured at Harvard between 1880 and 1912, and over 95 percent of whom have now died (Figure 7.6).

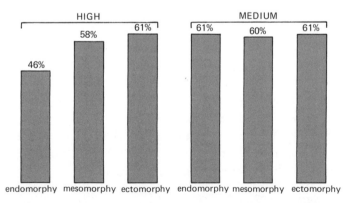

7.6 The percentage of men reaching age seventy by somatotype (dominant body form. The study is of Harvard men born between 1853 and 1894. *High* and *medium* refer to the degree to which the subjects conformed to the definition of the group to which they were assigned (3 and 4, medium; 5–7 high on a 7-point scale).

The men of average height and weight in college lived longer than the very tall, very heavy, or very short and very light. This suggests stabilizing selection.

We then looked at the children of those men in the group who had married. Again, men of average height had more children than the very short or very tall. We looked further, and found the same relationship in twenty-three hundred middle-aged factory workers in Chicago, seven hundred veterans in Boston, and one hundred eighty-nine tribesmen in the Solomon Islands. We looked through the medical literature and found confirmation in three unlikely sources. Newborn children of average birth weight have a considerably lower mortality than the very light or the very heavy. In work among sixty-two hundred Polish Army recruits, the number of living sibs, and hence family size, was greatest in men of average head form; it fell off in both the long-headed and broad-headed. And finally, among twenty-three thousand persons discharged from twelve hundred hospitals in this country during 1969, patients whose weights were near the national average for their age and height had shorter hospital stays, by two to three days, than those considerably underweight or overweight. So in respect to length of life, number of children, and duration of hospital stay, the individual of average physique is better off than the extremes.

Genetically, heterozygotes occupy the mid-portion of the distribution curve. When they are reproductively favored, the species keeps its present adaptation while perserving the source of variability for any future changes in environment.

Relaxed Selection. In the case of man, relaxed selection is implied in what I have said above about artificial selection; the term is used to refer to interference with natural selection through human culture. The force of selection for or against certain genes no longer holds in new environments, and the result of relaxed selection is a greater frequency of certain minor defects in agricultural or industrialized societies than in simple tribal groups. People in most tribal societies have decidedly fewer eye defects—colorblindness, nearsightedness, and astigmatism—than do industrial societies, which can compensate for some of such deficiencies by providing eyeglasses. So individuals with really poor eyesight are still not at a disadvantage for either survival or reproduction, and their deficiency genes, as increased by mutation or by unfavorable recombinations, are not selected against significantly. More than half of the population of the United States wears prescription eyeglasses, and in Japan the proportion is greater still. Some uncivilized people are better able to taste certain bitter substances, notably PTC (phenylthiocarbamide), although the nature of the selection here is not clear and the data are not of the best. They also have fewer deformities of the nasal bones and cartilage, that is, deviations of the nasal septum. It has also been suggested that the thinner skull of modern man compared to his immediate predecessors, like Neanderthal man, reflects less need to withstand blows on the head, and that his teeth, jaws, and face are all smaller because, even twenty thousand years ago, he began making more efficient stone tools.

Relaxed selection becomes much more serious when the genes in question are for major disease rather than minor functional disability. We can now keep persons with diabetes, cystic fibrosis, phenylketonunia, and other potentially lethal conditions alive to reproductive age. These genes will spread slowly in the population and will pose a burden on society for treatment. In case of social breakdown, as in war or natural disaster, such persons are extremely vulnerable.

SUGGESTED READINGS

Bajema, C. J. 1971. *Natural selection in human populations: The measurement of ongoing genetic evolution in contemporary societies.* New York: Wiley.

Bleibtreu, Hermann, ed. 1969. *Evolutionary anthropology.* Boston: Allyn and Bacon.

Cavalli-Sforza, L. L., and Bodmer, W. F. 1971. *The genetics of human populations.* San Francisco: Freeman.

Dobzhansky, T. 1964. *Mankind evolving.* New Haven: Yale University Press.

Dunn, L. C. 1959. *Heredity and evolution in human populations.* Cambridge, Mass.: Harvard University Press.

Moore, J. A. 1963. *Heredity and development.* Cambridge, Eng.: Oxford University Press.

Roberts, D. F., and Harrison, G. A., eds. 1959. *Natural selection in human populations.* New York: Pergamon Press.

8
HUMAN CHANGES AND RESPONSES

Of Mice and Men

We have been looking heretofore at the textbook operation, so to speak, of evolution in man. That is, we have dealt to a great extent with natural selection, mutation, or the founder effect as all these things affect flies, mice, and people. The important point is that man is indeed an animal, subject to these forces. But people are not mice. Mice have never changed their environment as radically as man has changed his own, and as he continues to do ever faster. Mice may have inhabited new lands by accidentally floating to them on natural rafts or blown-down trees, but mice have not made boats to go back and forth and back again; they did not come to America from Europe and then fetch African mice to pick their cotton. They do not X ray each other, nor put vitamin supplements in their own food. So human life has other dimensions. Man generates forces which can strongly affect his own biology, perhaps producing genetic changes of consequence but also having marked effects on the phenotype.

I shall have a good deal to say later on about what we have been doing to our own environment, and the number of problems this is posing. As to the human population itself, it has been exhibiting certain major trends away from its early days. First, it has become exceedingly mobile: not only have large segments of once regional peoples moved out over the world (Figure 8.1), like Europeans or Chinese, but individuals increasingly move about from place to place or country to city. Second, at least in the developing world, the environment is becoming more alike for

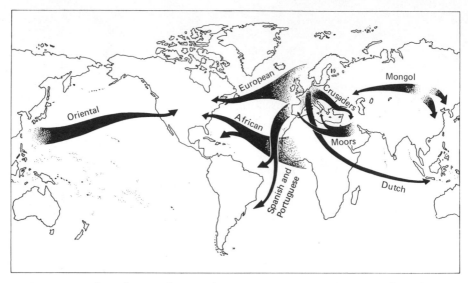

8.1 Some of man's major historical migrations. Mass movements such as these and others have helped stir the genetic pot, or even replaced it in others.

all peoples, even in such things as food, clothing, and shelter, as modern technology spreads everywhere; cities, unfortunately, become more alike in their special problems; medicine, fortunately, also reaches out and eradicates smallpox from the whole world. Third, people mix genetically at a far greater rate than when societies were parochial. This does not primarily mean race mixture, although that has gone on wherever the new travelers met the old inhabitants. Perhaps more important is the breaking up of local isolation within the same population, so that mating, which used to be highly local—practically the girl or boy next door—will now often bring together people from different parts of a country, pooling genetic resources more widely for whatever that may bring.

These are general ideas and trends. This chapter, however, is about some phenomena which are not fully explained, but which are striking examples of how we have been biologically changed in recent generations. These phenomena are increasing body size and earlier age of maturation.

Increasing Body Size

A change over time in some standard measure like body size is called a secular trend. For at least the past one hundred years, and probably longer than that, people have been getting bigger, and also getting bigger earlier. This increased adult size has been noted in Europe, the United States, and Japan. Adult height has increased at the rate of one inch per generation of thirty-odd years, so that men are now three inches taller than men in the 1860s. Moreover, men now reach their adult height in their very early twenties, instead of in their very late twenties, as they did one hundred years ago.

Among fathers and sons at Harvard between 1880 and 1920 (see Figure 6.3), Bowles found that the sons measured 1.3 inches taller than their fathers on entry to college, although the sons were a year younger. Between 1920 and 1946, American soldiers in World War II were 0.7 inch taller than those in World War I (Figure 8.2). Since 1946, there has been a further increase of 0.5 inch, making a total gain in height of 2.5 inches among U.S. men between 1880 and the present time. If this trend continues, we may produce a nation of basketball players, but fortunately the end is in sight. In fact, in a paper with precisely this optimistic title, "Increase in stature. Is the end in sight?" Bakwin and McLaughlin compared Harvard freshmen admitted in the 1930s with those admitted in 1958 and 1959. Boys from public schools were continuing to increase in height, whereas those from private schools were not. The implication was that the environmental conditions influencing growth, such as nutrition and medical care, were still improving for the public school boys, but had already reached an optimum for the private school boys, who were supposedly from more advantageous backgrounds in rearing.

My own work on Harvard men measured between 1870 and 1965

	World War I 1919	World War II 1946
numbers of men	97,000	25,000
age (years)	23	25
height (inches)	67.7	68.4
weight (pounds)	145	155

8.2 Body height and weight of U.S. males entering military service in 1919 and 1946.

confirms this possibility (see Figure 6.3). Among twelve families in which four generations had been measured at Harvard—father, sons, grandsons, eighty-five men in all—most of the height increase occurred between the fathers and sons (born 1858 to 1888), a minor portion between sons and grandsons (born 1888 to 1918), and practically none thereafter, among men born between 1918 and 1941. Further unpublished data of mine on six thousand men in two- and three-generation Harvard families show the same result. Again, the implication is that these favored families reached an optimum environment in the decade 1910–1919, so that the genetic determinants of height could reach their full expression early in this century. For families less well off, the environment is still improving, and with it the chance for complete expression of the genetic determinants of height.

Evidence that the secular increase in body size is slowing down comes from two more studies, Jeffrey Froehlich's among Japanese immigrants to Hawaii (Figure 8.3) and my own among mothers and daughters attending Mt. Holyoke and Wellesley Colleges.

Explanations. How can we account for the long-range increase in adult body size? Two general kinds of explanation have been advanced: the environmental, which is certainly true, and the genetic one of heterosis, or hybrid vigor, which may possibly be true as well. Needless to say, these two explanations, environmental and genetic, are not mutually exclusive; both factors can be operating. Clearly, major changes in physique within a single group over one or two generations, as among the Japanese in Hawaii, can only be environmental. The same reasoning applies, incidentally, to the epidemic increase of certain diseases in this country during the present century, such as coronary heart disease, peptic ulcer, and lung cancer. Genetic influences, by and large, simply do not operate that rapidly; they usually take many generations. But there is one

	Issei born in Japan	Nisei born in Hawaii	Sensei born in Hawaii
number	32	40	58
birth date	1890	1923	1945
age (years)	77	43	22
height (inches)	61.7	65.6	65.5

8.3 Secular trend in height among Japanese men in Hawaii: immigrants (Issei), first generation (Nisei), and second generation (Sensei) born in Hawaii. (After Jeffrey W. Froehlich.)

kind of genetic influence that could act within a single generation for a given group, and for several generations in a large national population. That influence is heterosis, described in the last chapter. It is the effect of larger, healthier, and more fertile offspring when two different inbred strains are crossed, although the evidence for men is slight—see below. This refers to strains. (Crosses of a more radical nature, especially in plants, may produce hybrids larger and more healthy, but sterile.) Now the relation of size to reproductive fitness is questionable. Fitness, strictly speaking, refers to the number of offspring reaching reproductive age. Nevertheless, the fact remains that increased body size does on the whole accompany fitness in the strict sense, in plants, animals, and possibly man. Gunnar Dahlberg, a Swedish geneticist, was the first to propose, in 1942, that heterosis might account for some of the secular increase in height. The theory is that increasing social and geographic mobility breaks down breeding isolates, a process that presumably accelerated with the invention of the bicycle.

Evidence for the environmental factor in stature increase is overwhelming. The Harvard fathers and sons are a good example: keeping heredity more or less constant, their stature increase over time reflects environmental change. To take another example, I have studied two hundred second-generation Italian-American workers in a single factory near Boston, all born and raised in the United States by parents who were themselves born near Naples, Italy. Within this homogeneous group, there was a steady increase in height as one compared men in their fifties with those in their forties, thirties, and twenties. The total amount of height gained over three decades was 2.1 inches. So large an increase over so short a period within a single group can only be environmental. The two specific factors in the environment that are mainly responsible are better nutrition and prevention or control of the diseases, chiefly infectious, that could retard growth in childhood. After the long bones stop growing, very little growth is possible, taking place mainly in the spine, so that adult malnutrition or disease will have negligible effects on growth.

Evidence for Heterosis. There are some slight signs that heterosis may occur in human populations. Frederick Hulse compared the height of Swiss migrants to California, their children born in California, and villagers who remained in Switzerland. The total height increase in the California-born men over the Swiss who stayed at home was 1.5 inches. Of this amount, 1 inch was due to the California environment and 0.5 inch to the fact that these men were more likely than those who remained in Switzerland to have had parents from different Swiss villages. Actually, no heterosis due to exogamy was demonstrable among the sedentes (stay-at-homes) or the migrants. However, among the Italian-American factory workers, men whose parents came from different villages averaged 0.75 inch taller than those whose parents were from the same village. Both of these sets of possitive results are at borderline levels of statistical significance.

There is also evidence, from inbred regions of France and from the Hutterite religious isolates, that inbreeding in man is associated with short stature—a roundabout way of suggesting that outbreeding, or hybridization, should be associated with tall stature. The trouble is, when one examines large populations of widely differing groups that have crossed, such as Caucasians with Chinese or Japanese in Hawaii, and when one examines such direct aspects of hybrid vigor as rates of stillbirth and neonatal mortality, one finds *no* evidence of heterosis. Even in stature, most descendants of racial crossing on a large scale are intermediate between the two parent stocks, rather than being taller than either one. We must conclude then that the case for heterosis is unproved. It remains just a possibility, pending further research.

Earlier Maturation

Soldiers in the Civil War continued to grow into their late twenties; nowadays, male growth is virtually complete by age twenty-two, and female growth by age eighteen. At all ages, children now are larger than they were even one generation ago; the *amount* of this difference of course decreases the closer the boys get to age twenty-two and girls to age eighteen. But several lines of evidence show that the speeding up of growth is not enough by itself to explain the increase in adult body size; the two phenomena are distinct.

The most dramatic evidence of the speedup in the maturation of girls is the age of menarche, or first menstruation. Progressive lowering of the age at menarche has occurred in the United States and Western Europe for at least the past hundred years, and in Eastern Europe since the late nineteenth century, having fallen in developed countries to an age between twelve and thirteen. The amount of acceleration has been between three and four months per decade, or roughly one year per generation, so that menarcheal age is now about three years earlier than it was a century ago. In general, city girls experience menarche earlier than country girls (Figure 8.4); the well-to-do, earlier than the poor; and American girls, earlier than their contemporaries abroad. Among immigrants to America

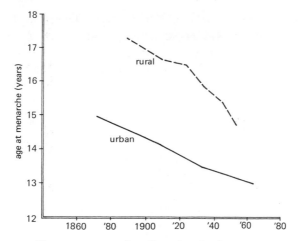

8.4 The comparable age at menarche of rural and urban American girls.

from Europe and Japan, the first generation of girls born in the United States has earlier menarche than their relatives who remain behind.

The best way to measure the acceleration would be to observe the age at menarche of successive generations within families, just as height and weight were observed within the Harvard families. This was done for sixty-six mothers and seventy-eight daughters (Figure 8.5), the daughters being subjects in a longitudinal study of child growth conducted at the Harvard School of Public Health between 1930 and 1956. (A longitudinal

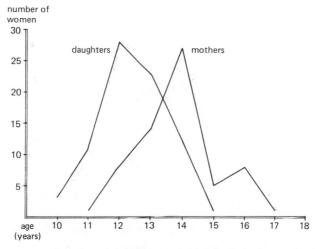

8.5 Age at menarche in sixty-six white mothers and seventy-eight daughters, 1930–1956. A study conducted by the Harvard School of Public Health. (Data, the author's.)

study is one in which the same individuals are followed through growth, as contrasted with a cross-sectional study, of different age groups measured at the same time; these groups may have other reasons than growth alone for differing.) On the average, the mothers were born in 1907 and their daughters in 1934. There was a striking acceleration of 1.5 years between mothers' and daughters' age at menarche, from 14.4 to 12.9 years. This was not due to a few very early maturers among the daughters; their whole distribution was moved ahead.

What can account for a difference of this magnitude in a single generation? Since the genetic factor was held more or less constant by comparing mothers and daughters, genetics cannot account for the acceleration. Age at menarche does have a genetic component; twins, sisters, and mothers and daughters are more alike in age at menarche than unrelated women. But in the present case we must clearly look to the environment.

Since the mothers were recruited from a hospital clinic, their socioeconomic status was low to moderate. Their daughters rose on the socioeconomic scale, as judged by comparing education of mothers and daughters, and education and occupation of the corresponding sets of husbands. But what specific factors in their environment could be responsible for their earlier menarche? Nutrition is the best-established influence on the rate of maturation. The amount of its effect is uncertain. The best estimate in the United States is based on white girls in Albama, where the mean difference in age at menarche between undernourished rural girls and well-nourished city girls was two-thirds of a year.

In the Harvard study, the daughters did have a better diet than the mothers, but the mothers were never as actually undernourished as were the Alabama girls. Even if we assume the extreme case—that the mothers as a group were as undernourished as the poor Alabama girls—nutrition would account for only 0.66 year of the total 1.5-year advance in age at menarche from mother to daughter. What can account for the remaining 0.8 year?

Medical advances could accelerate menarche by preventing disease. The evidence is, however, that most childhood disases are totally without effect on the growth of well-nourished children. Persistent chronic disease among malnourished children could indeed retard menarche, but this combination had not occurred among the mothers to any significant extent.

Psychosocial Influences. We are left with a large fraction of the secular advance in age at menarche to explain. The psychosocial factor is poorly understood and has been disputed. We have seen that urban girls begin menstruation earlier than rural girls—and urbanization is one feature of the environment that has increased steadily since the secular trend in age at menarche has been observed. Rural girls as well as urban girls show the secular advance. If so, urbanization might be interpreted as some unidentified aspect of the environment which is greater in the city than in the country, and which has been increasing over time in both city and country. In fact, in countries such as England and Holland, where there is now little difference between rural and urban ways of life, there is very little difference between the ages at menarche of rural and urban girls.

What are the effects of city life? Urbanization increases crowding and all kinds of stimulation—sensory, social, and intellectual. Urbanization increases competition and stress, while decreasing physical exertion. Another factor, coeducation, has also been increasing with time, but has been shown (in Sweden) not to affect age at menarche.

Still another possible factor is obesity, or body fat content. In the United States obesity is for the most part more urban than rural, at least in males; and fat girls generally have an earlier menarche than thin ones. Indeed there is evidence of a critical threshold of fat in the total body weight for the start of menarche, so that a girl fattens somewhat before menarche; furthermore an older woman on a stringent diet to lose weight may stop menstruating, only to resume the function when she adds fat again. In this view the acceleration of menarche in urban girls, which has been known since the Renaissance in Europe, may be related to their relatively greater obesity.

We might speculate that the effect of urbanization on age at menarche is comparable to the changes in wild rats brought about by domestication in the laboratory environment. Among the differences, sex glands develop earlier in the laboratory rat than in the wild rat. In recent experiments on the age of first ovulation in mice, Steven Vandenbergh et alia attributed 47.3 percent of the total variance to psychosocial factors (the male presence), and only 4.8 percent to dietary protein. Body weight, or rather, fat content, an hypothesized determinant of human menarche, had no influence on the onset of ovulation in mice. Incidentally, Vandenbergh et alia confirmed our own finding that physical growth and sexual maturation are separate processes. In short, better nutrition, improved health, and some as yet unidentified aspects of the material and psychosocial environment are all at work to accelerate development and lower the age at menarche.

Like the secular increase in height, the acceleration of menarche has slowed down among the well-to-do. Among 500 mothers and their 522 daughters entering Mt. Holyoke and Wellesley Colleges between 1930 and 1970, reported age at menarche was identical, 13.1 years, for mothers and daughters. Whatever its explanation, the secular trend toward earlier maturation is indeed, as the English growth authority, James Tanner, has said, "one of the most considerable phenomena of human biology at present, [with] a host of medical, educational, and sociological effects." To mention only a few: earlier reproduction, family formation, and divorce; dropouts from school; lowered voting age; and increasing tension among young people who chafe against rules formulated for earlier generations who were less mature physically and physiologically at comparable ages. Whether intellectual and emotional maturity have kept pace is another matter.

At the other end of the female reproductive span, the age at menopause, when the menstrual periods stop, there is some doubt as to whether this has been occurring later in life. One generation ago most authorities gave forty-five as the average age in the United States, but some put it later. All current studies place the time around fifty years of age. Menopause is more difficult to time precisely than menarche since it occurs over a period rather than at a definite time. Whatever the situation for individual women, there has been no rise in births to women over forty-five years of age, making it dubious whether the possible retardation in age at menopause has any evolutionary implication for the species.

SUGGESTED READINGS

Baker, Paul T., and Weiner, J. S., eds. 1967. 2d ed. *The biology of human adaptability*. New York: Clarendon Press.

Birdsell, Joseph B. 1975. *Human evolution: An introduction to the new physical anthropology*. New York: Rand McNally College Publications.

Damon, Albert, ed. 1975. *Physiological anthropology*. New York: Oxford University Press.

Dubos, R. 1967. *Man adapting*. New Haven: Yale University Press.

Harrison, G. A., Weiner, J. S., Tanner, J. M., and Barnicot, N. A. 1964. *Human biology*. New York: Oxford University Press.

9
BEHAVIOR GENETICS
By Winthrop A. Burr

Behavior genetics is not fundamentally different from other genetics except that behavioral characteristics are often harder to define and measure than physical ones. Behavior geneticists conduct inquiries into the nature of encoded genetic information and the ways in which this information interacts with environmental factors in the development of behavioral repertoires. Since in man we usually cannot do the elegant experiments necessary to build complete pictures of these complex interactions, we may be forced to deal with behavior genetics problems in more superficial ways. For example, we are often reduced to asking the extent to which a behavioral characteristic is environmentally versus genetically determined. In other words, we are obliged by ignorance of detailed mechanisms to grapple with questions about human intelligence, personality, ability, and mental illness in the old and oversimplified framework of heredity versus environment, or the nature-nurture argument.

Nature of the Evidence

The study of behavior genetics is potentially extremely important from an evolutionary point of view since, in many nonhuman social species, relatively small changes in individual behavior are often accompanied by large changes in social organization which, in turn, have major effects on a species' adaptedness to a given environmental niche. Thus the evolution of behavioral characteristics (which is only possible to the extent that they are genetically determined) as opposed to the evolution of morphological characteristics, constitutes one of the most rapid means available to a species for responding to new environmental pressures.

Although it is sometimes assumed that behavior genetics is, at base, the genetics of brain structure and function, it is important to realize that gene-environment interactions involving behavior may occur when the genes involved have their primary effects in organs other than the brain. Two illustrations of this point are: alcohol consumption in the rat and the illness phenylketonuria (PKU) in man.

Alcohol Consumption. In experiments with rats, alcohol consumption by the animals is found to be partially genetically determined in that there are significant strain differences that breed true. Strains which drink more alcohol have, on a genetic basis, higher levels of the enzyme which catalyzes the first step in the metabolic breakdown of alcohol. These strains may also differ in brain structure and function, but at least in terms of present evidence, it appears that the higher consumption is more probably related to a high rate of alcohol metabolism. Thus the behavior may be chiefly a reflection of liver function.

Although the human genetic evidence is contradictory, similarities in alcohol tolerance and drinking habits can be measured in adopted children, comparing them both with their biological parents and with their adoptive parents. This research does indicate the operation of a genetic factor in human reactions to alcohol and drinking habits. Other research has pointed to racial differences in vasodilatation (flushing) following small doses of alcohol to newborns. Furthermore, some ethnic groups are more likely to flush after drinking alcohol than others. This phenomenon is probably more common in the Japanese than in the people of European descent. A major problem in the study of human alcoholism is that since chronic heavy alcohol consumption induces so many biochemical changes, it is difficult to establish any as having existed prior to the development of the drinking. Rat alcohol drinking is probably not a good model for human alcoholism since environmental stress decreases the rat's consumption instead of increasing it.

PKU. Human PKU provides another example of how a gene (or lack thereof) may influence behavior through an organ other than the brain. PKU is a rare illness which occurs in the absence of the enzyme (phenylalanine hydroxylase) necessary to convert the dietary amino acid phenylalanine to tyrosine. Individuals with this illness are homozygous for a recessive allele which is defective in that the required enzyme fails to be produced by cells in the liver. The result is a buildup of phenylalanine in the blood. Phenylalanine is transported into the brain from the blood stream in competition with certain other amino acids. The profound mental retardation that occurs may be the result either of a direct toxic effect of excess phenylalanine or of a lack of other amino acids displaced by the phenylalanine. The picture is made complex by the fact that some individuals are known who have very high phenylalanine levels without mental retardation. In any case, profound behavioral effects result from the absence of a normal liver function.

In man, most of the behaviors studied so far have complex causes, and it seems probable that polygenic systems must be involved. In certain lower animals, however, rather complex behaviors result from the operation of one cr a small number of genes. In these cases, the effects of the presence or absence of one or a few genes are quite subtle and specific. For example, certain strains of honeybee are resistant to a common infectious disease, American foulbrood, on a behavioral basis. Resistant hives protect themselves by cleaning the hive of dead larvae and thus inhibiting disease transmission. Genetic analysis of this behavior tentatively indicates that it is probably controlled by two genes which assort independently; that is, they appear at separate loci in individuals. One gene controls uncapping the cells where the dead larvae are; the other controls removing the dead larvae from the hive.

How a single DNA sequence can have such complex effects is, of course, anyone's guess at the moment, but the chances are that the gene concerned does not contain information for the entire behavior pattern, but rather for some small part of it which is necessary as a trigger for the rest; that is, the ability to tell whether larvae are dead or alive.

Intelligence

In moving from these relatively simple considerations, we now enter a region of greater relevance but of considerably lesser clarity: human intelligence. The subject of human intelligence has such profound social, emotional, and moral overtones that scientific consideration of it has, at times, been nearly impossible. In no other field has the nature-nurture controversy remained so heatedly alive. Intelligence, whatever it is, is one of the most highly valued human attributes. The fact that some may possess it in a high degree, through no effort of their own, while others are denied its benefits, is contrary to our ideas of fairness. Thus it is usually considered more socially useful to deemphasize any hereditary component in intelligence because to do otherwise might merely discourage efforts to improve education for the less favored. It might, in fact, discourage some of the less favored themselves. The underlying assumption is that what is influenced by genes is not subject to the effects of environment, an assumption that needs careful examination. We have some obvious examples of how it can be wrong.

The genetic constitution for PKU determined severe mental retardation until it was discovered that elimination of phenylalanine from the diet during the years that the brain is developing could result in normal or near-normal intelligence.

Down's Syndrome. This syndrome (already described in Chapter 5) is a genetic condition caused by an extra dose of most or all of chromosome number 21. Mental retardation, one of the many features of this syndrome, is so severe that most affected individuals used to spend their lives in institutions. Recently, however, intensive educational efforts, combined with keeping the children at home during the early years, have enabled almost all to lead relatively independent, productive lives. Down's syndrome is no longer usually considered an acceptable reason for admission to institutions for the retarded in Massachusetts. In this case, what appeared at one time to be an upper limit to intellectual capacity set by the genes was modified by intervention, such that there is now a new higher upper limit. This in no way weakens the argument that Down's syndrome is a genetic condition, or that the condition affects learning ability.

IQ. In man, the most objective measure we have of intelligence is the IQ test. Whether this measures all that we mean when we use the word "intelligence" is an unresolved and possibly unresolvable question. IQ test results are powerful predictors of school performance, even when teachers do not know the test results and therefore are not influenced by them. School performance has importance in determining the acquisition of certain credentials (high school diploma, college admission, graduate school admission, Phi Beta Kappa, etc.). These credentials, in turn, have a determining effect on occupational success and prestige. Thus it is that IQ is related to success in life. It is not at all clear, however, that within groups holding similar credentials, IQ test scores have any relationship to success. The probability of becoming a physician is related to IQ, success as a physician in comparison with other physicians (measured in academic, monetary, or other terms) is determined to some extent by medical school attended and this, in turn, is related to school performance and IQ; but among graduates of a single medical school, IQ test results bear little, if any, relation to success. The same statement could be made for other areas of graduate and professional training. This may mean that IQ as we measure it reflects a narrower range of abilities than we sometimes assume, and that success depends heavily on other things, such as energy, creativity, interpersonal skills, etc. In fact, because of the heavy emphasis placed by credential-granting bodies on abilities related to IQ, our society may be hampering development of a certain number of otherwise promising people. On the other hand, it is important to bear in mind that holders of similar credentials occupy a narrow range of the IQ spectrum and the above results may simply reflect the fact that within narrow limits IQ is not important, even though it is important within broader ones.

IQ is an attribute with a high heritability component in individuals. It reaches a stable value at about seven years of age and, although 70 percent of the population shows changes of up to ten points in score at different times, it usually remains stable within those limits until old age. The volume of studies demonstrating the heritability of IQ in different countries, at different times using a variety of methods, places this finding above serious question. It is not a simple matter, however, to decide *how* heritable IQ is, or—more importantly—how measures of heritability are to be properly understood. Before proceeding to untangle some of these issues, let us look at some data.

Correlation Coefficients. Correlation coefficients are statistical measures of the closeness of relationships betwen two variables. They are bounded by the values −1 and +1, each of which indicates a perfect relationship (all points falling on a single straight line) between the variables; that is, the value of one variable can be predicted exactly from the value of the other. If the sign is negative, the value of one variable gets bigger as the other gets smaller, while if the sign is positive, both variables increase or decrease in tandem. A correlation coefficient of zero indicates no relationship at all; that is, knowledge of the value of one variable in no way restricts the possibilities for the value of the other one. There are numerous ways of trying to understand the meaning of intermediate correlation coefficients. For purposes of this discussion, it is enough to look at them in comparison to each other.

Twin Studies. Consider now some representative correlation coefficients between members of the following kinds of pairs. Monozygotic (MZ) twins reared together are correlated in terms of their IQ scores at about 0.90. MZ twins reared apart correlate somewhat less closely (0.75–0.85). Parents and children (0.50), parents with each other (0.50), sibling pairs reared together (0.50–0.55), and sibling pairs reared apart (0.47–0.49) are considerably less closely related in terms of IQ. Unrelated

people reared together (0.24–0.27) and unrelated people reared apart
(−0.01–0.00) have the most distant relationships. (See Figure 9.1.) Note
that although being reared apart does attenuate the size of the correlation,
it does so much less dramatically than decreasing degrees of genetic relationship between members of pairs. In fact adoption studies have shown
that adopted children's IQs are only slightly less closely related to their biological parents' IQs than are ordinary parents and children, while the
relationship between adopted children's IQs and those of their adopted
parents is on the order of that of any other unrelated pair living together
(Figure 9.2).

So much for the bald facts. To achieve a useful understanding of

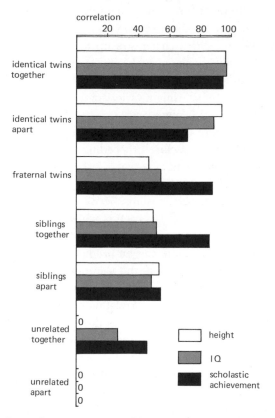

9.1 The correlation between twins, siblings, and unrelated children and their
height, IQ, and scholastic achievement. Height is included as a standard of comparison in association drawn from purely physical features. (SOURCE: Herrnstein,
R. J. 1973. *I.Q. in the meritocracy.* Boston: Atlantic Monthly Press, p. 161.) Note,
Herrnstein adapted his chart from Jensen, Arthur R. 1969. Reducing the Hered-
ity-Environment Uncertainty: A Reply. *Harvard Educational Review,* Summer,
39: 349–383.)

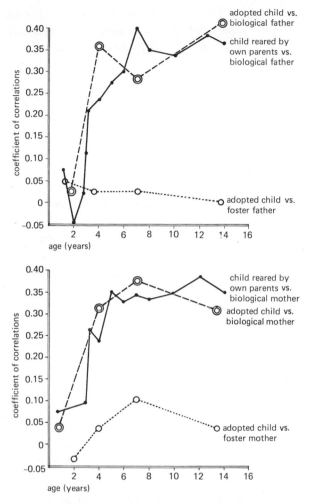

9.2 Correlation between the IQ of children and the educational level of real and foster parents. (From Honzik, M. P., *Child Development*, **28**, 1957.)

them will require consideration of some methodological issues, The most elegant methods we have for studying the operation of genetic factors in man involve the use of twins. Monozygotic (*MZ*) or one-egged twins are genetically identical individuals. Dizygotic (*DZ*) or two-egged twins are genetically separate individuals, no more alike genetically than any other pair of siblings. On the average, groups of sibling pairs share half their genetic material, but any one pair of sibs may theoretically share almost any fraction thereof.

The cornerstone of the twin method is the comparison of same-sexed *DZ* twin pairs and *MZ* twin pairs in terms of differences between members of the pairs. MZ twins differ because of environmental differences between them (if one ignores the possibility of somatic mutation). DZ twins differ on both environmental and hereditary bases. The twin method rests heavily on an assumption that MZ and same-sexed DZ twins share their environments to the same extent; that is, twins are as likely to have had similar or different experiences regardless of whether they are of the MZ or same-sexed DZ variety. If this were so, then any greater similarity on the average between members of MZ pairs and members of DZ pairs would reflect the operation of genetic influence because environmental factors would tend to move pair members closer together or further apart to an equal degree in both MZ and DZ groups.

This assumption, unfortunately, is not wholly defensible. Environmental noncomparability for MZ and DZ twins can begin in the uterus. DZ twins always have separate placentas and separate fetal membranes. MZ twins, on the other hand, may have a variety of possible intrauterine relationships, depending on when the developing embryos split (Figure 9.3). (Siamese twins are MZ twins in whom the developing embryos never split completely.) MZ twins may share all their membranes and have a single placenta, or they may be completely separate as are DZ twins.

There is often an inequality in the distribution of material blood supply when two individuals share a uterus. This inequality may have the effect of producing relative stunting in the growth of the less-favored member of the twin pair. This inequality is less likely to be experienced by ordinary siblings who each have the whole of the maternal blood supply to themselves. The degree of inequality may depend on the relationship of membranes and placentas and may differ substantially between MZ and DZ groups. Specifically, MZ twins are less likely than DZ twins to have had an even division and do differ more than DZ twins in birth weight. This effect, operating in a twin study, would tend spuriously to reduce the calculated genetic contribution. Postnatally the environments of MZ and DZ twins may differ in their degree of similarity as well. In the behavioral sphere it has been observed that MZ twins have more friends in common, play together more, and are more often responded to similarly by others than are DZ twins. Insofar as these factors would produce more similar development for MZ twins, they would have the effect of spuriously elevating the apparent genetic contribution.

Adoption Studies. The other mainstay of human genetic research is the adoption study. The usual strategy here is to compare children with their biologic families and with their adoptive families. If the relationship between the adoptees and their biologic families is stronger than the one between the adoptees and their adoptive families, there is evidence that

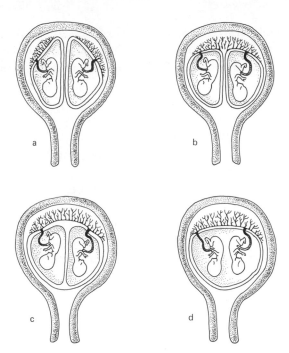

9.3 Schematic drawing of various possible twin pregnancies in the uterus. (a) Monozygotic or dizygotic twins with separate amnions (water sac containing the embryo), choria (outer membrane enclosing the embryo), and placentas. (b) Monozygotic or dizygotic twins with separate amnions and choria. (c) Monozygotic twins with separate amnions and single chorion and placenta. (d) Monozygotic twins with single amnion, chorion, and placenta. (Redrawn from Potter, E. L. 1948. *Fundamentals of human reproduction*. New York: McGraw-Hill; Stern, C. 1973. *Principles of human genetics, 3rd ed.* San Francisco: Freeman.)

genetic factors are more important than environmental ones. The major assumption is that children and their biologic families have not shared their environment to a degree that might outweigh the environmental contribution of the adoptive family. This may not be the case. All children have experienced the maternal uterine environment and it is a rare study indeed that can claim all subjects were removed from their biologic families at the time of birth.

Another type of problem results from the fact that adoption agencies tend to match biological and adoptive families to some extent. This has the effect of providing the adoptee with an environment that has a better-than-random chance of being like the one in his biologic family. This means that, to some extent, whatever relationship is observed between the adoptee and a biologic family may in fact be the product of a similar or shared environment. Matching for things such as physical appearance may produce a certain degree of genetic matching also so that the adoptee and his adoptive family may share more of their genetic material than would be expected by chance alone. These confounding factors can have complex hard-to-predict effects on the study results. Insofar as intrauterine environment or early childhood experience are important in determining the characteristic under study, adoption studies are likely to lead to overestimates of the importance of genetic factors.

The final conclusion must be that although twin studies and adoption studies are the closest natural analogs of more fully controlled animal work involving large purebred lines, they have their defects and must be taken with a grain of salt.

Heritability. Genetic studies usually have as their aim the calculation of heritability. While a detailed arithmetic discussion of heritability is not relevant here, it is important to have some understanding of what heritability actually measures.

In describing a group, the degree to which its members differ from each other is expressed by a statistical quantity called the variance, which measures the amount of dispersion in the sample from the average value for the sample. The more homogenous a group is, the lower its variance. In heritability calculations, the total variance is treated as an entity to be partitioned according to its various sources. The percentage of the variance that would disappear if all the subjects had the same genotype is defined as the heritability.

The sources of the total phenotypic variance are: variance in genotype, variance in environmental forces, and a third factor which depends on interactions between genes and environment. [Figure 9.4.] (Gene-gene interactions and gene-environment correlations can also contribute, but are beyond the scope of this discussion.) In other words, differences between members of a group result from the fact that their genes differ, their environments differ, and the particular interactions between their genes and environments may itself produce differences between them. The interaction factor can create thorny problems in calculating heritability. It is worth illustrating by an example. A group of children with Down's syndrome raised in an institution and a group of genetically normal children raised at home would differ in IQ. The difference would be produced by the fact that the children differed genetically by the fact that one group had been brought up in an institution while the other had not, and by something else. Children with impaired learning potential suffer greater intellectual deficits as a result of institutionalization than normal children do. Therefore the interaction between the genetic constitution for Down's syndrome and the environment of the institution produces

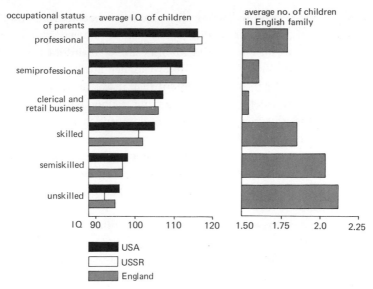

9.4 Occupation, IQ, and family size, a composite picture drawn from several sources. The IQs of children from all three countries decrease in proportion to the socioeconomic status of the family.

greater impairment than would occur through a summation of the genetic liability (Down's syndrome in a normal environment) plus the environmental liability (normal child in an institution). The interaction itself is thus a separate source of variance that must be taken into account. The problem that interaction creates in heritability calculations is that with the exception of a few extreme cases such as the one illustrated, we are hardpressed to know whether or not interaction is operating and how important it is.

Since heritability is a fraction of the total variance, it is subject to influence by the environmental fraction and the interaction fraction. Thus, for example, if the genetic contribution to the variance is held constant while the environmental contribution is varied, the measured heritability will change also.

Perhaps an example from agricultural genetics will help make the point. If a handful of genetically heterogeneous seeds are sown in a uniform type of soil and all are exposed to the same conditions of light and water, the resulting plants will differ in height. If we assume there are no interaction effects, the differences will be due to the genetic differences between the seeds and the measured heritability of plant height will be high. If, on the other hand, the same seeds are sown on several different kinds of soil and if light and water also vary from seed to seed, then the differences in height will not be due just to genetic differences, but also to environmental ones (again ignoring the possibility of interaction.) Heritability, which is the fraction of the total variance attributable to genetic variance, will be lower in this experiment than it was in the first one, as a matter of simple arithmetic, because the other fraction has become larger. Bear in mind that this happens without in any way changing the genetic constitution of the plants involved.

Heritability is a population concept, not an individual one. It says nothing directly about individual history or the degree to which an individual's intelligence might have been modified had his life circumstances been different.

Understanding these concepts may go part of the way toward explaining one of the chief paradoxes of research into the genetics of human intelligence. Adoption studies and twin studies point to a high heritability of intelligence as measured by IQ tests. Reasonable past estimates run between 60 percent and 80 percent. When this group of findings, which seem to indicate high heritability of IQ, is juxtaposed to findings that racial and social groups differ in IQ, it is tempting to conclude that the differences must reflect different genetic potentials. (The mean IQ for blacks in the United States has often been reported as approximately 15 points lower than that for whites, and these reports have been the basic source of controversy. The average correlation between IQ and social class runs between 0.4 and 0.6.) There are other possible explanations, however, which should be considered. First, for reasons outlined above, high heritability does not preclude high sensitivity to environmental influence (Figure 9.5). The racial differences may partly reflect the operation of greater environmental variance in cross-racial comparisons than in adoption and twin studies. Environmental variance would be decreased by matching between biological and adoptive families. Many of the twin series have come from quite homogeneous groups. However, a study comparing black and white twins in Philadelphia did show higher heritability in the white group. Apparent heritability is likely to be smaller in those studies where the environmental component is enlarged; that is, when the environmental factors important for IQ development are particularly large, as between biological and adoptive families or between twins of a pair. This is not to argue that anything would reduce the heritability of IQ to an artifact. The point simply is that the measurement of heritability is uncertain and the concept itself tells much less about genetic potential than we would like to know.

Cross-cultural and cross-racial comparisons also raise questions about

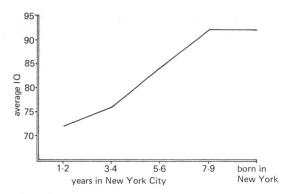

9.5 IQ and environment. The IQ of twelve-year-old Southern-born black girls related to the number of years residence in New York City. According to research worker Otto Klineberg who collected the data in the 1930s, scores improved proportionately to the number of years the subject had lived in the North (Klineberg, O. 1935. *Negro intelligence and selective migration.* New York: Columbia University Press; 1944. *Characteristics of the American Negro.* New York: Harper.)

how adequately IQ tests distinguish between ability on the one hand and achievement and the effects of opportunity on the other (see Figure 9.5). They are validated to some extent by their heritabilities. However, they are all performance tests, and learning plays a part in any performance above the level of simple reflexes. In order to get around this problem, considerable effort has been devoted to develop a culture-free test, with only very limited success. Many feel now that the effort is in vain. The problem may never be solved until we are able to define intelligence in neuroanatomical or neurophysiological terms. If we are ever able to do this, we may find that IQ as we presently measure it rests on a different physiological and different genetic basis in different individuals or different groups. If this were the case, it would be one more reason for heritability to vary from group to group and study to study.

There is a tendency to think of IQ as a single attribute, although we more reasonably assume it to be a complex resultant of many semi-independent neurophysiological processes. In this sense, IQ must be thought of as a polygenetically determined characteristic. For an analogy, you might think back to the example of skin color—also a polygenetically determined characteristic with a continuous distribution. Skin color too is a complex resultant of many semi-independent processes: malanosome morphology, melanocyte activity and distribution, epidermal thickness, and others.

IQ level can certainly be diminished in any of a host of ways. Mental retardation appears as a product of many chromosomal anomalies of over a hundred single gene defects, and also results from a long list of infectious physical, chemical, psychological, and social injuries. Although there are many specific causes known, approximately 80 percent of cases occur in the absence of any specifically determinable etiology.

Personality

This area is perhaps less emotionally loaded than that of intelligence, since the relationship between success and personality is less specifically defined, and since racial comparisons have not been much of an issue. On the other hand, the field is even more confused because there has been great difficulty designing valid and reliable measures of personality and emotional characteristics.

Validity and reliability are statistical attributes which tests must have before they can be useful for research. The degree of validity is the degree to which the test actually measures what we think it measures. One of the most venerable psychological tests, the Minnesota Multiphasic Personality Inventory (MMPI), which is a self-report questionnaire, is routinely analyzed in terms of empirically derived scales developed to discriminate between various patient and nonpatient groups in Minnesota in the 1940s. The names of the scales ("schizophrenia," "psychopathetic deviate," "masculinity-femininity") reflect the original uses to which they were put. More recent research, however, has shown that several of the scales are not valid indicators of what they were originally alleged to measure. For example, experience with the test has shown that individual scores on the "masculinity-femininity" scale reflects something quite different from what we ordinarily mean by masculinity or femininity. For one thing, male students typically get high femininity scores, which indicates that the scale may really measure something more closely related to intellectual interest. Thus the MMPI, still a useful test in many contexts, has had to be revalidated as an instrument of personality description, by comparing various scale profile configurations with assessments gathered in other ways, such as extensive interviews. In this process, scale names have become superfluous and have been dropped by many workers.

The chief validity problem in psychological tests stems from the fact that the constructs (depression, anxiety, passivity, ego strength, etc.) suffer from imprecise definition and cannot be objectively assessed. One is often in the unsavory position of having only someone's impression (either the subject, observer, or both) to compare the test results with. To get

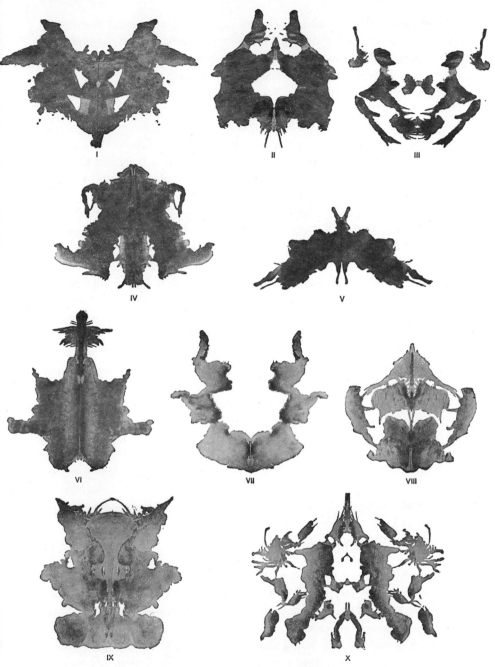

9.6 The standard Rorschach test. The individual inkblots, some of which are in color, are presented to the subject one at a time. Individuals are asked to describe what they perceive in the pictures. From the tone and nature of the responses, the examiner is able to gain some insight into the personality of the person taking the test.

around this, a considerable amount of work and some recent slight progress has been aimed at defining objective physiological correlates of emotion. Serum lactate as a measure of anxiety, salivary gland secretion as a
measure of depression, and electronic recording of the activity of the
muscles of facial expression are three relatively current areas of research
interest.

The degree of reliability in the tests is the degree to which differences in tests results from time to time and from observer to observer
reflect real changes in the subject as opposed to slippage in the measurement. One of the most spectacular of psychological tests, the Rorschach or inkblot test (Figure 9.6), is capable of yielding
extraordinary subtlety and depth. But it is plagued by low reliability
because interpretation depends so much on the individual tester's ability to make intuitive leaps and because, remarkably enough, a well-
standardized scoring system has never been developed.

The problems of validity and reliability are, to a degree, intertwined.
Although a test could have perfect reliability and no validity, the converse
is not true. Defects in reliability must impair the validity of any test.

Despite difficulties with the basic measurements, a relatively large
amount of research has been carried out on the inheritance of personality
characteristics. As you might imagine, the degree of consistency and comparability between studies is not overwhelmingly high. A very tentative
conclusion can be drawn that a tendency to introversion versus extrover-
sion may have higher heritability than other personality characteristics.
Measured heritabilities here are, however, considerably lower than they
are with IQ.

Neonatal Behavior
and Cultural Practices

The behavior of newborns varies considerably from one to another. This is
manifested by differences in irritability, sleep-wake cycles, alertness, and
so forth. These differences, like all biological differences, are probably the
result of gene-environment interactions. The environmental factors
operate prenatally and during the birth process. In normal pregnancies,
these probably do not constitute major sources of variation, although abnormal intrauterine conditions have profound effects. Normal variations in
intrauterine environment have not been well-studied, for obvious reasons.
Prenatal environmental effects on behavior have been shown in rats born

to mothers stressed during pregnancy with a loud, unpredictable buzzer. Such rats, when raised by adopted mothers, showed greater emotionality compared to similarly adopted rats whose biological mothers were unstressed. Emotionality in the rat is measured by the amount of defecation which occurs in unfamiliar, open surroundings. Although the validity of this measure is open to question, and extrapolation of the findings to man requires daring and imagination to say the least, it does demonstrate that prenatal environmental variables can affect behavior.

There has been much speculation about subtle prenatal behavioral effects in man. It has been shown that certain analgesic and anesthetic drugs given during labor can have an effect on newborn behavior which persist for days. It has been suggested that a less-responsive newborn may be less effective at eliciting certain maternal behaviors and that the mother-child relationship may thus be permanently altered, even by an agent that has transient effects on the newborn. In support of this interpretation, there is some evidence that events in the first few days of life can have relatively long-lasting effects on mother-child relationships, at least in some circumstances. It was found that in a group of unwed adolescent mothers, allowing the infant to room-in with the mother during the first few days of life was associated with greater maternal interest in, and ease with, the infant several months later.

Nonetheless, it does seem probable that in the absence of any disturbing influence, differences in behavior at the time of birth are, to a large extent, genetically determined. On the average, the newborns of different racial and ethnic groups have been observed to differ, and these differences can interact in interesting ways with child-rearing practices. An example is provided by the Zincanteco Indians of Mexico. Pediatricians studying Zincanteco newborns observed considerable differences between them and North American white babies. Zincanteco babies were as alert and responsive and spent as much time awake as North American babies, but were quieter, cried less, and showed less motor activity. The differences were apparent from the earliest days of life, and were reinforced by the practice of carrying babies tightly swaddled, with their faces covered most of the time. Babies were also breast fed at the earliest sign of restlessness and were not allowed to cry for any length of time. In later months, they were not stimulated nearly as much as North American babies and their exploratory activities were not particularly encouraged. On the average, they tended to become quieter, less exploratory children. Their development in terms of motor milestones (sitting, grasping, walking, talking, etc.) was parallel to North American babies, albeit about a month delayed through the first year. On certain developmental tests, the

Zincanteco showed considerable facility in imitation and less facility in initiation. The authors concluded that this quiet imitative personality was adaptive where social cohesion and conformity to norms was important for group survival, as it was to the Zincanteco and only is to a lesser extent among ourselves. The point of the example is to show that congenital behavioral differences, quite possibly genetically determined, seem to be interacting with cultural variables in an adaptive manner. The story lacks detail and contains some speculative elements, but it provides hints as to how we may approach an understanding of human behavior genetics from an evolutionary perspective.

Schizophrenia

There has been recent progress in understanding the genetics of schizophrenia, which is the most common of the severe mental illnesses. Schizophrenia is a term used to delineate a group of psychiatric patients who exhibit bizarre thinking and behavior, usually onsetting in late adolescence or young adulthood and usually resulting in major disruption of personal relationships and ability to work or study. Definitions of the illness are often vague and unsatisfactory, reflecting the difficulty of capturing the essence of the disorder in words. The meaning of the term schizophrenia ("split mind") bears little relationship to current thinking about the syndrome. Diagnosis of schizophrenia has always been a distressingly subjective matter, with the result that the boundaries of the disorder have expanded and contracted with different observers and changing diagnostic fashion. This has played havoc with attempts to do research and has markedly limited the comparability of studies. Recently though, through the use of standardized structured interviews combined with computerized data handling techniques, it has been possible to make the diagnosis with a reliability comparable to other clinical and pathological assessments. Past definitional difficulties notwithstanding, there is agreement that schizophrenia is common; so common that approximately 1 percent of the U.S. population can be expected to be so diagnosed at some time in their lives.

Recent adoption and twin studies, some of which have taken advantage of the extensive case registers and adoption records available in certain Scandinavian countries, have shown that inheritance plays a major role in the development of schizophrenia.

Adoption studies have shown that the risk of developing the disorder depends more heavily on its presence or absence in the biological family than in the adoptive one. The risk is the same whether mother or father was schizophrenic, which argues against a nongenetic prenatal effect. *MZ* twins have concordance rates generally estimated over 60 percent while *DZ* twins have concordance rates of 20–30 percent.

In addition, biological relatives of schizophrenic patients have higher rates of personality disturbances which, although not as severe as schizophrenia itself, have many features characteristic of it and have been called schizophrenia spectrum disorders. In all likelihood, a polygenic system is involved, with the fullblown disorder requiring a relatively high dose of deleterious genes. Smaller doses produce lesser degrees of deviancy or, in some cases, might be adaptive. In this vein, some investigators have found the relatives of schizophrenic persons to be more creative than the average. Since the disorder is common, and since the fertility of affected people has, at least until recently, been considerably reduced owing to decreased marriage rates and long periods of institutionalization, the speculation has been put forward that a balanced polymorphism is involved. In other words, the frequency of the disorder might be kept high by increased reproductive fitness among persons with smaller numbers of the relevant alleles. Unfortunately for this idea, it has never been shown that the relatives of schizophrenics are more reproductively fit than the average, and there are no particularly coherent suggestions currently available as to what the selective advantage might be.

Criminal Behavior

The idea that criminality is a form of genetic degeneracy is an old one. It had, however, been dismissed as a serious possibility until the mid-1960s when workers in England discovered, in a group of institutionalized men, a few tall individuals with histories of violent crime, who had an *XYY* karyotype. This observation was repeated at other institutions and the idea gained wide currency that the *XYY* karyotype was associated with tallness and violence (see Chapter 5). The introduction of *XYY* karyotype evidence played a role in at least two murder trials. The picture became somewhat blurred, however, when studies of normal populations also turned up *XYY* individuals. As the situation now stands, it appears that *XYY* karyotypes are overrepresented among violent populations, as are *XXY* (Klinefelter's syndrome) and, more strikingly, *XXYY* karyotypes. However, most *XYY* individuals are neither violent nor tall. No one can say what the risk is that any given XYY individual will develop unfavorably.

A prospective study, designed to follow babies with XYY and other abnormal karyotypes from birth, with the aim of assessing the risk of developing behavioral or perceptual disorders and providing early intervention in the form of special education and counseling, has been under way until recently. This study was brought to a halt by critics who charged that merely discovering such individuals and informing their parents of a possibly increased risk would create unwarranted expectations that might develop into self-fulfilling prophecies. Although outside reviewers did not agree that potential risks to the children outweighed potential benefits to them, the principal investigator ended this project citing the harassing nature of the criticisms to which he had been subject.

This illustrates quite nicely that human behavior genetics, however great its methodological difficulties, and however unclear its results may sometimes be, suffers from no lack of current relevance or ability to fire the emotion as well as the intellect.

SUGGESTED READINGS

Herrnstein, Richard J. 1973. *I.Q. and the meritocracy.* Boston: Atlantic Monthly Press.

Loehlin, John C., Lindzey, Gardner, and Spuhler, J. N. 1975. *Race differences in intelligence.* San Francisco: Freeman.

Rosenthal, David. 1970. *Genetic theory and abnormal behavior.* New York: McGraw-Hill.

Rosenthal, David, and Kety, Seymour S., 1969. *The transmission of schizophrenia.* New York: Pergamon Press.

Wilson, Edward O. 1975. *Sociobiology.* Cambridge, Mass.: Harvard University Press.

10
AGE: THE HUMAN LIFE CYCLE

So far we have been concerned with the background and mechanisms of human evolution: how man arose; and the factors, cultural as well as biological, which have operated in human differentiation. From this point we consider the biology and ecology of modern man: where we are now, and where we are going. All of evolution did not take place long ago and far away. It continues today around us, in both civilized and tribal societies, although some rates and some selective pressures, for different characteristics, may be changing.

We first consider the variation or diversity of human beings along four main axes: age, sex, race, and body build. We shall look mainly at variation in the patterns of disease and of physique that are associated with each of the above axes. There are other important kinds of difference, both among individuals and among populations, in such things as blood groups and serum proteins, but these are dealt with in standard treatises in anthropology and will not be covered here. I shall merely remark that such traits support the other physical, linguistic, and archaeological evidence for human differentation. In other words, the major groupings of man, although overlapping, have their special attributes, whether one classifies by geographic origin, by physical appearance, by biochemical makeup, or by patterns of susceptibility to disease.

Disease and Age

Let us begin by viewing disease as an agency of natural selection, and the impact of age on this. Disease has surely been one of the major mechanisms of human evolution. Persons unable to cope with the rugged conditions of primitive life are soon eliminated, removing their genes from the next generation. A striking example is the fate of the albino in the tropics. A child with any serious congenital disease or disability can hardly survive. Even if kept alive, the child is most unlikely to reproduce, and thus his or her genes are lost. Other examples would be the universal presence of bacterial and parasitic infections in existing tribal communities. A child (or for that matter an adult) who could not handle such infections, for genetic reasons, would soon be eliminated. On a worldwide scale, whole populations in the Pacific and the Americas have been drastically reduced in numbers—or even eliminated entirely—by the introduction of diseases to which no immunity had been built up by generations of previous exposure. It has been estimated that millions of American Indians died from smallpox, which was first introduced by one single member of Columbus's crew. Smallpox was the leading factor in colonizing New England—far more potent than gunpowder, alcohol, or the Bible, although all of these helped. Measles and venereal disease performed the same grisly function in the Pacific islands. Measles is still a lethal disease among unexposed populations, whether Eskimos or Melanesians, although it is a mild childhood disease in Europe, the United States, and Japan. By the same token, tuberculosis is less prevalent and less severe among people who have developed resistance, by virtue of residence in cities for generations, than is the case among primitive and rural people previously unexposed.

If disease is to act as a selective factor in evolution, it must of course affect reproduction—that is, it must occur before or during reproductive life. After reproduction has ceased, natural selection acting on the individual can no longer affect the makeup of the next generation.

To illustrate these points, consider mortality and morbidity in the Solomon Islands and in the United States, in the years 1900 and 1963 (Table 10.1). The Solomon Islands can be taken as an example of primitive people for whom we have some reliable statistics. There was, for 1963 in the Solomons, a preponderance of infectious diseases causing morbidity, which means illness in general. In contrast, the United States in that year had no infectious disease among the seven leading causes of morbidity. As for mortality, the leading causes of death in the Solomons were again infectious diseases. The picture in the United States in 1900 was roughly similar, with a few chronic diseases, such as heart disease, cancer, and strokes, trailing behind; whereas in 1963, with the conquest of the infectious diseases, there was only one type of infectious disease, influenza plus pneumonia, among the leading causes of death. Incidentally, the importance of accidents as a leading cause of death seems to be similar in primitive and civilized man, even though ours occur on roads at high speeds, and theirs do not. It would seem that millenia of selection against fools and the accident-prone have not been very successful!

We may next contrast the leading causes of death among children,

Table 10.1. MORBIDITY AND MORTALITY IN THE SOLOMON ISLANDS AND IN THE UNITED STATES [a]

MORBIDITY

Solomons [b]	United States [c]
Malaria	Heart conditions
Accidents	Arthritis and rheumatism
Skin diseases	Mental and nervous conditions
Respiratory diseases	Impairment (except paralysis) of back and spine
Gastro-intestinal	Hypertension without heart involvement
Eye disease	Visual impairment
Acute infection	Asthma—hayfever

MORTALITY

Solomons: 1963	United States: 1900	United States: 1963
Malaria	Influenza, pneumonia	Heart disease
Respiratory disease, acute	Tuberculosis	Cancer
Gastrointestinal, acute	Gastro-enteritis	Strokes
Epidemics, various	Heart disease	Accidents—motor vehicle
Accidents	Strokes	Accidents—nonmotor vehicle
	Chronic nephritis	Influenza, pneumonia
	Accidents	Diseases of early infancy (newborn)
	Cancer	Congenital malformations
	Certain diseases of early infancy (newborn)	Diabetes mellitus
	Diphtheria	Cirrhosis of liver
		Suicide

[a] In 1973, heart disease, cancer, and strokes were still the first three causes of death. Diseases of infancy dropped to ninth place. Congenital malformations left the list and was replaced by emphysema and asthma.

[b] Based on hospital admissions and outpatients.

[c] Causing limitation of activities.

Table 10.2. LEADING CAUSES OF DEATH IN CHILDREN
AGES ONE TO FOURTEEN IN 1935
AND JUST THIRTY YEARS LATER
(RATE PER 100,000) [a]

1935		1965	
1. Accidents	39.0	1. Accidents	22.6
2. Pneumonia	33.8	2. Cancer [b]	7.1
3. Diarrhea and Enteritis	12.3	3. Congenital malformations	5.0
4. Appendicitis	10.2	4. Influenza and pneumonia	4.8
5. Tuberculosis	10.0	5. Meningitis	1.0
6. Diphtheria	9.8	6. Gastritis	0.9
7. Influenza	9.7	7. Diseases of the heart	0.8
8. Diseases of the heart	8.9	8. Homicide	0.8
9. Measles	7.9	9. Vascular lesions (stroke)	0.7
10. Scarlet Fever	6.0	10. Cerebral Palsy	0.6
11. Whooping cough	5.5	11. Cystic Fibrosis	0.6
12. Cancer [b]	5.0	12. Nephritis	0.6

[a] SOURCE: *Vital Statistics of the United States.*

[b] Includes leukemia and Hodgkin's disease.

Table 10.3. INFANT (UNDER ONE YEAR)
DEATHS PER THOUSAND LIVE BIRTHS IN
SOME SELECTED COUNTRIES;
DATA 1972 [a]

Denmark	14	Guatemala	83
Spain	16	Jordan	99
Great Britain	18	Cambodia	120
United States	19	Nicaragua	123
El Salvador	53	India	131
Mexico	61	Malawi	174
Malaysia	75	Nigeria	180
Equador	75	Angola	203
Colombia	76		

[a] SOURCE: *World Population 1973*, U.S. Department of Commerce Social and Economic Statistics Administration, Bureau of the Census.

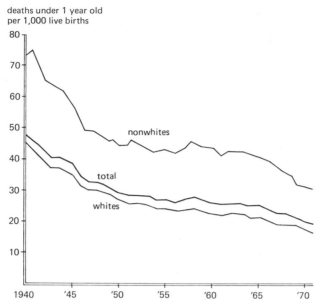

10.1 U.S. infant mortality rates 1940–1971 by race. (From *Social Indicators,* a publication of the Social and Economic Statistics Administration, U.S. Department of Commerce, 1973. Note that terms *white* and *nonwhite* are those used by the U.S. Bureau of the Census.)

10.2 A Pacific island paradise, but with disease rates, death rates, and infant death rates significantly higher than heavily industrialized countries.

aged one to fourteen years, in the United States in 1935 and, just thirty years later, in 1965 (Table 10.2). In the preantibiotic era, infections predominated. Today they are much less common, and genetic diseases and congenital conditions which reflect the prenatal environment are much more prominent causes of death. In contrast to primitive societies, or in contrast to modern underdeveloped countries where one out of three or four children or even higher percentages die before reaching maturity, the death rates in the United States for children from all causes are extremely low (Table 10.3 and Figure 10.1). For all twelve leading causes of death among children, the combined death rate is now less than 50 per 100,000, or 0.5 per 1,000. This is the current situation in the advanced, industrialized countries of Europe, North America, and Japan. In the thirty years between 1935 and 1965, we reduced childhood mortality rate from 1.5 to 0.5 per thousand. By way of contrast, the comparable death rates for children in underdeveloped (agricultural) countries like India and parts of Latin America are 250 to 300 per 1,000. In such undeveloped agricultural countries, one out of every three or four children dies between the ages of one and fifteen years (Table 10.4). Think about this the next time you are tempted to yearn for some bucolic Eden (Figure 10.2), without automobiles, telephones, or smog. In such an earthly paradise, one out of three of us would not even be here; he would have died before his fifteenth birthday (Figure 10.3). And the older you get, the more you depend on medical progress. (I myself would have died on six separate occasions, had it not been for modern medicine.) Our society may not guarantee happiness; no society can. But it can virtually guarantee an indispensable component of happiness—namely, life itself.

So much for the diseases of childhood. Remember that we are considering disease in relation to age. How about the diseases of middle life? Young adults have low death rates in all societies. For males, the leading cause of death is accidents. For females in uncivilized societies, childbirth is the main hazard, with infection the actual lethal agent. In advanced societies, the hazard of childbirth is slight, and the death rates for women in the reproductive years age are below those of men, reversing the situation among primitive groups.

In later adult life, among ourselves the well-publicized major killers take over—namely, heart disease, cancer, and stroke. Since one must die of something, in the developed countries these diseases have now replaced infections. But among primitive peoples, infections are still preeminent.

Death rates in late adult life are not very much better among the civilized countries (Figure 10.4). In fact, in the United States since 1900, there has been very little lengthening of life, after the age of forty-five. It is true that the life expectancy of a newborn child in this country has increased from forty-five years in 1900 to seventy years at present. However, all but a few years of this increase are attributable to the fact that fewer children now die, not to longer life after maturity.

It is obvious that, just as age can determine the disease pattern

Table 10.4. MORTALITY RATES (EXPRESSED AS PERCENTAGES) FOR CHILDREN FROM BIRTH TO VARIOUS AGES [a]

Ethnic Group	Country	Percentage
Moï	South Vietnam	30 infant mortality (under one year of age)
Various	West Irian (New Guinea)	10–35 infant mortality; 41–58 before the age 15.
Navaho	United States	13.9 infant mortality
Various	Uganda	25 infant mortality 30 before age 5
	Imesi W. Nigeria	29.5 infant mortality; 69.7 before age 5
Various (Kha and Xos, Phou Thai, Primitive Seks, and Primitive Nhos)	Central Laos	26.1–35.3 before age 18
Gola	Liberia	80 conceptuses die before 18 years
Various	Malaya	28.7–59.5 die before their mothers
Kapaukus	West Irian (New Guinea) Highlands	38.7–47.2 die before their mothers
Muruts	North Borneo	33.8 die before their mothers

[a] SOURCE: Polunin, I. V. "Health and diseases in contemporary primitive societies," in *Diseases in antiquity*, Brothwell, D. E., and Sandison, A. T., eds. Springfield Ill.: Thomas, 1967.

The data were compiled by Polunin from a variety of anthropological and medical sources dating between 1939 and 1963.

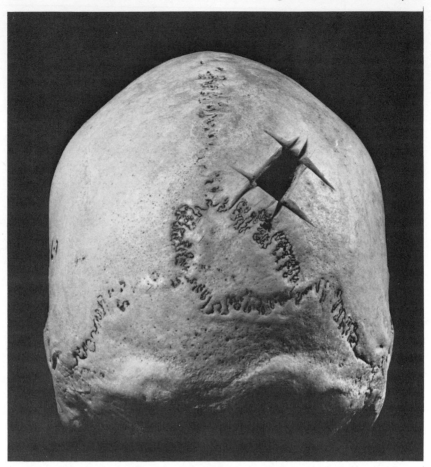

10.3   A failure of folk medicine. Trepanning (creating a hole in the skull by scrap-
ing, drilling, or cutting) was a surprisingly widespread practice in prehistory
which, in Melanesia and some African societies, has continued to the present. This
violent operation sometimes killed the patient (as happened here) but often did
not (as indicated in some skulls by the subsequent growth of bone tissue). Trepanning probably was resorted to in order to relieve pressure on the brain and to
cure recurring migrain headaches. (courtesy Peabody Museum, Harvard University)

exhibited by a society, so in turn the disease pattern of a society helps
shape its age composition. For example, a group with high childhood
mortality will show many infants and relatively few adults. On the other
hand, an advanced society with low infant and adult mortality will have a
much more balanced percentage of persons under age fifteen and over age
forty-five. In fact, these percentages are practically equal in the United
States today (Figure 10.5).

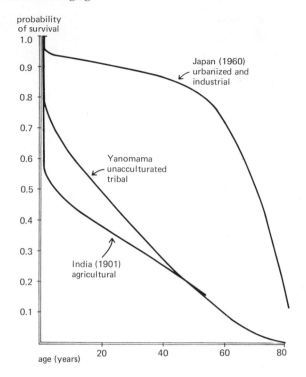

10.4   Life curves for three types of population. (The Yanomama are South American Indians.) (Reprinted by permission of Dr. James V. Neel.)

Just as mortality shapes the age structure of a population, so the age structure influences breeding patterns and hence the genetics of the small groups in which man has lived for 99 percent of his time on this planet. The two main points of the above discussion are: (1) the importance of infectious disease for most of human evolution (note that malaria is still the most common worldwide disease); (2) the fact that any influences on evolution, including disease, must exert their effect before or during the reproductive years.

Physical Growth and Aging

Let us now turn to normal physical growth and aging. The first point to make is methodological. What we want to know is how the individual grows and ages. The best way to find this out is by a so-called longitudinal

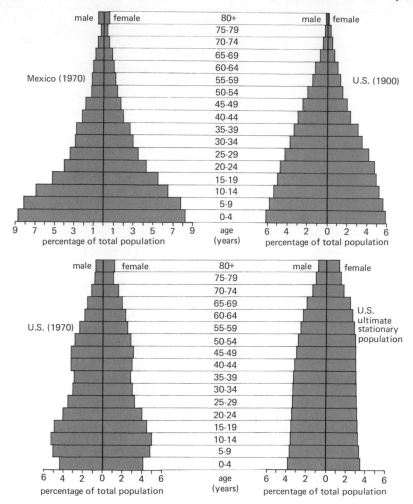

10.5 Birth and death rates give characteristic shapes to age distribution pyramids in different populations. Mexico and, to a lesser extent, the United States (1900) are characteristic of fast-growing populations with high birth and death rates where the life expectancy is relatively low. As birth rates decline, the pyramids become more columnar. (Redrawn from Freedman, Ronald, and Berelson, Bernard. 1974. The human population. *Scientific American*. 231:38. And from Westoff, Charles F. 1974. The populations of the developed countries. *Scientific American*. 231:114, 116.)

study, in which the same individual is observed at frequent intervals for a lifetime. The practical problem of doing this for any significant number of persons are staggering, in terms of cost, effort, and time required—particularly when the subjects outlive the investigator. Nevertheless, several such studies have been carried out on several hundred subjects, from birth to age eighteen, and a few of them are being continued into adult life. At the other end of the life span, there are now perhaps twenty longitudinal studies in adult life, of which the largest has been operating for twenty years—a sample of the town of Framingham, Massachusetts. Obviously, only longitudinal studies can give rates of growth, development, or aging.

Nevertheless, much of our knowledge of the human life cycle, particularly in adult life, still rests on the much easier cross-sectional study, in which large numbers of subjects of different ages are observed at one time. It is assumed that the individual will, over time, follow the course, say from age thirty to sixty, which corresponds to the average differences between persons in the population aged thirty and sixty at the present moment. But a little thought will show that several other explanations are possible, in addition to the one we are looking for: age change in the individual. Men of sixty are shorter than men of thirty. Individuals do shrink late in life for three reasons: bone loss from the vertebrae, or bones of the spine; kyphosis (forward bending of the thoracic spine); and, chiefly, from shrinkage of the intervertebral disks. But in a cross-sectional survey, we may be seeing several other things in addition to individual shrinkage with age. We may be seeing selective survival—that is, short men may actually live longer than tall ones, so the older men, who are the survivors of a cohort born at the same time, will be shorter than those who died off; and thus, on the average, shorter than the generation following them. A second possibility is selective migration of tall men out of the population under study. In actual studies, migrants do tend to be taller than people who stay put. And finally, as you know, there has been a secular or long-term trend toward increasing height in the population. This means that men who are now thirty are taller than men now sixty when the latter were themselves thirty years old. As a result of the secular trend, younger men would be taller than older ones, even if shrinkage with age did not occur.

The net result of all these confounding factors is that differences between age groups, as found in across-sectional study, do not necessarily reflect age changes in the individual. In fact, the few studies we have show virtually no difference in height among individuals measured in their early twenties and again in their mid-fifties. So the shrinkage in height must begin later. The same is true for grip strength and lung capacity: there is no difference between a man's score in his twenties and in his fifties. The moral is: Accept no substitutes for longitudinal research on the human life cycle.

Types of Growth. Figure 10.6 shows four growth curves of different

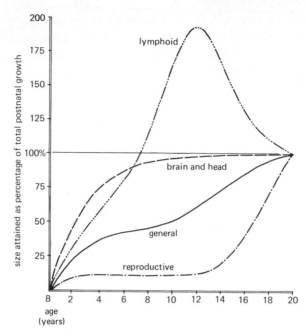

10.6 Size growth and maturity of body systems attained as a percentage of total postnatal growth.

tissues and different parts of the body. The horizontal scale is years; the vertical scale represents percentage of adult size reached, 100 percent being adult size. Most skeletal and muscular dimensions, including height and weight, follow the general curve. So also do organs such as the liver, spleen, and kidneys. This curve is gradual at first, with a marked rise beginning around age twelve, which is called the adolescent spurt. The reproductive organs, both internal and external, follow an exaggeration of this curve. The prepubertal phase is lower, and the adolescent spurt much more marked. The reason for the difference in the two curves, the general and the reproductive, is that the skeleton, muscles, and most organs are more sensitive to certain hormones affecting growth, namely those of the pituitary and thyroid glands; whereas the reproductive organs are more sensitive to the sex hormones, the androgens and estrogens, which increase markedly at adolescence.

The brain, together with the skull, the eyes, and the ears, develops earlier than any other part of the body. This curve seems to decelerate smoothly after birth, although recent very careful studies suggest the existence of regional growth irregularities in the brain before adolescence. At any age during childhood, the head in general is more advanced than the remainder of the body, and the top portions, namely the eyes and the brain, are more advanced than the lower portion, namely the face and jaw. Babies seem to have big eyes. This is because the rest of their faces are small at first, while the eyes are close to their adult size.

The fourth type of growth, that for the lymphoid tissue of tonsils, adenoids, appendix, and lymph nodes, follows a quite different growth curve. Lymphoid tissue reaches its peak before adolescence and then, probably under the direct influence of the sex hormones, declines to its adult value.

These and other differential growth rates are very often the mechanisms of morphological evolution. To give only two examples, man is distinct from apes in having longer legs relative to the body and arms (see Figure 2.14). This comes about because man (of course, as an evolutionary adaptation) has a greater relative velocity of leg growth than have the apes, from early fetal life onward. That is, relative to growth of the rest of the body, man's legs grow faster than an ape's from the very start. To give another example from the other end of the body, the joint between the head and the spine is farther forward on the skull in man than on other primates (see Figure 2.15). At birth, this distinction is not present: other primates have the joint just as far forward. Growth in the primate skull base is largely in front of the joint, pushing it back; whereas skull growth in man is equal in front of the joint and behind it. The joint therefore remains in the same place. Within the human species, individual and population differences in body size and shape result from such differential growth rates.

Obviously there must be controlling forces which synchronize these complex developments. A few such mechanisms are known.

1. One is the principle of *target-seeking* or *self-stabilizing*. Growth which has been interrupted by disease or malnutrition will catch up to its original path or trajectory after the disturbance is removed. This regulatory force has a genetic basis. It is stronger in females than in males.

2. Another principle is the existence of *maturity gradients* within regions of the body. For example, in both the arm and the leg, the part farthest from the body is, at all ages, farther along toward its final, mature value than the part closer to the body. Also, the head is always in advance of the trunk, and the trunk in advance of the limbs.

3. A third principle is *feedback regulation*. Secretion of the pituitary hormones affecting the various target glands—specifically, the thyroid, adrenals, and gonads, or sex glands—is adjusted to the level of hormones produced by these target glands. When the target gland secretes just the right amount of its hormone, the pituitary adjusts its output to maintain the equilibrium.

4. Rose Frisch and Roger Revelle, at the Harvard Population Center, have proposed a fourth mechanism regulating growth, the notion of a *fixed amount of body mass*, sensed by the body and setting in motion the events of adolescence. For girls in the larger-bodied human populations they propose that a body weight of 106 pounds ± 3 or so triggers off the end of the adolescent growth spurt and menarche. Differences between the age at menarche in different populations, they propose, mean that they take a longer or shorter time to reach the magic weight. A tentative finding of my own, that the length of reproductive span in women seems to be constant, may point to a similar kind of body sensing, this time of a built-in time tally. After a certain number of years, the ovaries no longer respond to pituitary stimulation, although there are plenty of ova remaining.

Post-Adolescent Growth. In man, unlike some other mammals such as the rodents, growth ceases when the epiphyses, or ends of the long bones, unite with the shaft. This occurs at the end of adolescence, when sex hormone production is at adult levels (Figure 10.7). But growth continues in bones and cartilage which grow by apposition (at the end surfaces, rather than between epiphyses and shaft). As a result, height increases between ages twenty and thirty by about one-fourth of an inch, on the average, due to growth at the tops and bottoms of the vertebral bodies. Growth of the clavicles (or collar bones) produces broader shoulders after thirty than at twenty. Head and face dimensions continue to grow very slowly up to age sixty. The long noses and ears of old people are not illusions but are real. In fact, adult overactivity of the pituitary gland, known as acromegaly, causes growth in the head, face, ears, hands, and feet, showing that these bones are capable of growth even after the long bones have stopped growing.

So much for the skeleton. As for weight, we all know what happens between the twenties and the fifties in an affluent society like our own. The average weight gain among American men is some fifteen pounds, and among women twenty-three pounds. Weight is lost again late in life, starting around age sixty. This weight gain does not occur in most primitive societies, which are less addicted to gluttony and sloth.

On a cross-sectional basis, most physical abilities, such as grip strength and lung capacity, reach a peak in the late twenties, then decline slowly to the fifties and sixties (Figure 10.7). Visual accommodation—that is, the ability to read print close to the eye—holds up fairly well until around fifty, the bifocal age. Hearing declines steadily, from a peak early in life.

Glands and Timing. The human growth curve is characteristic of

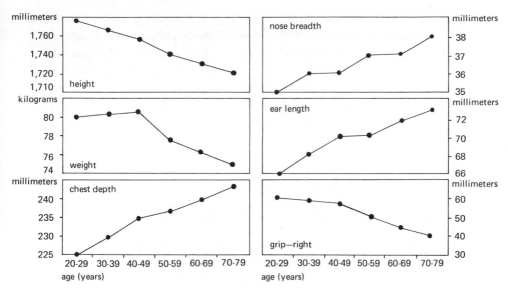

10.7 Some selected variables in the study of physique and age. (Study of 2015 white veterans of Boston, Massachusetts, made by the author.)

primates. It is shared by chimpanzees and rhesus monkeys but not, for example, by rodents or cattle. The mouse and rat grow steadily, with little interval between weaning and puberty, and no adolescent spurt except in their reproductive organs. The essential difference in growth between the rodent and the primate is delayed puberty, and the mechanism lies in the brain. The maturation of the hypothalamus, which initiates the events of the adolescent spurt, is delayed until the associative areas of the cerebral cortex are relatively advanced. This delay has been carried successively farther in monkeys, apes, and man. The evolutionary significance to these animals is enormous: it promotes learning and socialization, tending to inhibit intermale aggression, on the part of young, able, but still tractable males before they come into sexual competition with adult males. In baboons males are powerful fighters, and their maturation takes twice as long as that of females.

The factors influencing growth are both genetic and environmental, and of course these two sets of factors interact. The Y chromosome slows the rate of growth, both in normal males and in Klinefelter males, who have the XXY chromosome constitution. Men are taller than women because men grow more slowly and hence for a longer time. The rapid growth of women ceases when their epiphyses close, around age sixteen, whereas men continue to grow to age eighteen and beyond.

The endocrine glands are one of the chief mechanisms for translating the genetic instructions into the adult form at the pace permitted by the environment. Even before birth, some of the fetal endocrine glands are active—specifically, the thyroid, gonads or sex glands, and the adrenals. From birth to adolescence, the major gland controlling growth is the pituitary, through its growth hormone. Next in importance during this period is the thyroid. At puberty, the hormones secreted by the gonads, or sex glands, and by the adrenals take over.

Evolutionary Significance of Aging

Why is a long senescence possible, if natural selection has no use for an individual after he or she has finished reproduction? Old persons consume food and occupy space that would otherwise be available for their offspring. In general, most animals do not survive their reproductive period for very long. Selection has favored long-lived families in our own ancestors because old people served two socially useful purposes which favor group survival: first, they sat with the babies while parents were working, and second, they served as a store of accumulated knowledge which they transmitted through the spoken word. The situation has now changed. In our modern society, old people are still useful for minding the children, but their function as reservoirs of knowledge has been replaced by the printed word, the radio voice, or the televised picture. In fact, it has been argued, perhaps not seriously, that in a society like ours, where parents still have to assume the expense of rearing children, and where wealth can be inherited, there is now a form of selection favoring shorter lives. The argument runs that parents who die young, leaving their money to their children, will thereby permit their children to afford larger families than if each generation had to earn its own money or had to wait for its inheritance beyond its own reproductive period.

Moving from these flights of fancy back to solid biological ground, let us take a look at rates of aging. Like rates of growth, individuals, and organ systems within the individual, age at different rates. One of the best single clue to age is the amount of graying in the hair. In rare cases, hair will turn gray prematurely in persons otherwise quite young. Brains too are subject to varying degrees of premature senility. A famous story is told about the late Supreme Court Justice Oliver Wendell Holmes. He was walking with a friend one day, at the age of ninety-one. After passing a handsome girl, Holmes stopped and turned around to watch her. Turning back to his friend, he said with a sigh, "What I wouldn't give to be seventy again!" This introduces the notion of biological time, as distinct from calendar time. A wound or cut of the same length and depth will heal five times faster in young animals, including man, than in old ones. This is due to age changes in the collagen, which is the fibrous and elastic supporting tissue of the body. Loss of elastic tissue results in the thin, wrinkled skin of older persons. It takes much longer for an old person's skin to snap back after a fold has been lifted, than for a young person's skin. Wrinkling occurs mainly on skin exposed to sunlight, which destroys elastic tissue in skin.

    Factors Controlling Rate of Aging. Like the factors controlling the rate of growth, both genetic and environmental influences control the rate of aging. Nobody doubts the existence of a genetic biological clock in growth and development, although the precise timing and extent of growth may be impaired by adverse environmental effects such as malnutrition and disease. However, until very recently nobody was sure whether aging represents the running down of a biological clock or merely the accumulation of environmental insults, including radiation and disease. If the latter were true, we might hope to extend the human life span greatly by developing means of protection.

    As for the mechanisms believed responsible for aging, mammalian bodies contain three kinds of components. First are cells that continue to divide throughout life. These include most cells, although they divide at different rates. Second are cells that cannot divide or renew themselves at all, such as nerve cells. The only thing that can happen to them is to die. Third, there is noncellular material that may have much or little turnover, such as collagen (a structural protein) and intercellular substances. Accordingly, there have been three grand theories of aging, as follows: change in the properties of multiplying cells; loss of fixed cells; or change in the inert materials of the body. It now appears that there is indeed a biological clock and that aging and eventual death are programmed into the dividing cell. You may recall the famous experiment of Alexis Carrel, who thought he had demonstrated that the individual cell was immortal and could survive indefinitely in isolated tissue culture. For his most dramatic experiments, Carrel used fibroblasts, or connective-tissue cells, from the chick embryo. Some cells can so survive, if they are cancerous or if they have been transformed, whether spontaneously or by a virus. But normal cells cannot. Hayflick suspected that Carrel's nutritive medium was contaminated by fresh cells. When Hayflick used a filtered medium which presumably excluded such cells, he found that fibroblasts from the lung of a human fetus can divide only about fifty times before the cell line dies out. Fibroblasts from a human adult have far fewer divisions, and those from an old person still fewer. So it appears that cells have a finite life span, and that aging and death are programmed into the cell and, by analogy, into the organism. The human life span is finite, even if we could shield people from radiation, disease, and other onslaughts of the environment. It is thus unlikely that we can ever prolong the average human life span much beyond the biblical three score and ten. And with this sobering thought, we leave the topic of age and aging.

SUGGESTED READINGS

Comfort, A. 1964. *Aging, the biology of senescence.* London: Routledge.

Kohn, Robert R. 1972. *Principles of mammalian aging.* Englewood Cliffs, N.J.: Prentice-Hall.

Krogman, W. M. 1972. *Child growth.* Ann Arbor, Mich.: University of Michigan Press.

Rose, C. L. and Bell, B. 1971. *Predicting longevity:* Methodology and critique. Lexington, Mass.: Heath Lexington Books.

Tanner, J. M. 1962. *Growth at adolescence.* 2d ed. Springfield, Ill.: Thomas.

Young, J. Z. 1974. *An introduction to the study of man.* New York: Oxford University Press.

11
A DIFFERENCE
OF SEX

If the axis of age has infinite gradations, the axis of sex has only one, settled for good at the moment of conception. Still, that one familiar difference has wide and varying biological consequences; among other things, it interacts with age, as we shall see below.

Sex Ratio

The sex ratio is the relative proportion of males and females in a given population, or part of it. Almost never is this ratio 50:50 (Figure 11.1) for two reasons. First, more males than females are born. Second, at all ages—in fact, right through prenatal life—males have higher death rates, until the last ancient survivors are virtually all women.

A special note: The term *sex ratio* itself refers to a specific kind of figure. It would be logical to use relative percentages or proportions of the sexes, for instance, 51 percent males. But by usage, the sex ratio always gives males as a percentage of the females, that is, the number of males for every one hundred females of the same age (or all ages).

Sex Ratio at Birth. For the United States, the birth sex ratio has for a long period of time been 105 to 106 (males per 100 females), and a similar figure has held for most other countries having accurate records. In the United States, whites have a higher ratio than nonwhites: for the years 1922 to 1926 the white ratio was 106.6; the black ratio, 105.4. Similar black-white differences have been reported from Africa and elsewhere,

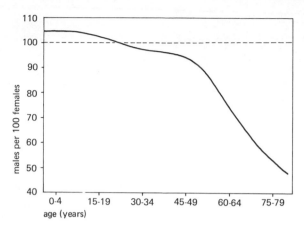

11.1 1960 data of human sex ratios by age in England and Wales. (From Lerner, I. Michael. 1968. *Heredity, evolution and society.* San Francisco: Freeman, p. 121, using data from A. S. Parkes.)

and they may represent a real biological difference. Some very high ratios of 110–115:100 have been reported from Korea, apparently well-documented. That is, they are based on good records for births in university hospitals and not on parents' reports, which might underestimate female births. These high Korean sex ratios have not been explained, but they are certainly far beyond those recorded for any other population.

If equal numbers of males and females were conceived, the predominance of males at birth could theoretically result from a greater prenatal mortality of females. This is not the case, however, except in early embryos; male rates for total intrauterine deaths and for still births exceed the rates for females. Recent developments in cytology, which we have already mentioned, have made it possible, with great reliability, to determine fetal sex from sex chromatin in the cell nucleus: females are chromatin-positive, males are chromatin-negative. Cells shed by the fetus can be obtained from the pregnant uterus and examined for Barr bodies. Also, while sex cannot be reliably determined by gross inspection before the fetus is three months old, cytological diagnosis is now possible at ten to twelve days. (At this early developmental stage, contrary to later fetal and postnatal ages, female embryos tend to die more often than males.) The best present estimate, based on the sex of naturally aborted fetuses, places the human sex ratio at conception around 120:100. This is the primary sex ratio; what we see at birth is the secondary sex ratio.

There is no doubt as to the fact that more males than females are conceived, but the reason is unknown. Presumably, it lies in some biochemical, physical, or physiological difference between the two kinds of sperm, those with an X and those with a Y chromosome, in the environment of the female reproductive tract. Shettles has made the plausible suggestion that the Y-bearing sperm, being slightly smaller, travels farther and faster, and so is more often successful in fertilization. The greater mortality of males than of females begins in the uterus and continues throughout life, and this is something encountered widely in the animal kingdom, not being confined to man. Its explanation is unknown, despite a great deal of speculation. There is no need to wonder whether our society is doing something that kills off little boys faster than little girls—or male fetuses faster than female fetuses, or adult males faster than adult females—whatever it is, all human societies and most animal societies are doing the same thing. Actually, in the United States, childhood environments up to about age five are fairly similar for boys and girls, but still the male death rate is higher. After five, when environments begin to differ, the excess male mortality becomes even greater. It reaches a special peak around age twenty, due in large measure to accidents, with suicide and homicide contributing appreciably.

Let us return to the sex ratio at birth. Any factor that acts to decrease mortality in the fetus and the newborn would obviously increase the sex ratio at birth, by tending to preserve the primary (conception) ratio. This might occur, for example, with improved health care, and lower prenatal mortality. It has in fact occurred among nonwhites in the United States; that is, with improved health care, the sex ratio at birth has been increasing among nonwhites, although it is still smaller than among whites.

Sex Ratio in Wartime. In addition to medical care, another environmental influence on the human sex ratio is the significant rise in the proportion of male births during wartime or shortly after wars. This has been noted in England, France, and Germany during World War I, and in the United States and Germany during World War II. MacMahon and Pugh in 1954 established this intriguing phenomenon for the United States and showed that it was not due to an increase in first births (which are also associated with a higher sex ratio) nor to the age of father or mother. The rise was quite small numerically, being from 105.8 to 106.1, but is based on millions of births. It is uncertain whether it was due to a higher sex ratio at conception or to a change in intrauterine environment, resulting in decreased male mortality. In fact, two authors concluded that if the increase in male births was not attributable to Divine Providence, they could think of no better explanation.

In 1958 Bernstein, without presuming to question the working of Divine Providence, suggested a genetic mechanism for its operation. He found that persons in the German *Who's Who*, marrying between 1900 and 1918, produced first-born offspring with a sex ratio of 124:100 for children born within eighteen months of the marriage date, and only 85:100 for those born after eighteen months. Bernstein concluded that wartime conditions, requiring the husband to be away for long periods of time, favor the early reproducers and thereby the production of sons.

Bernstein's hypothesis has recently been confirmed on large numbers of births in Finland and Australia. Among first-born children of many thousands of subjects, the closer the birth to the date of marriage (starting with nine months, so as not to include premarital conceptions), the more likely was the child to be a male. Renkonen believes that some mothers become sensitized to successive male fetuses and produce antibodies against them.

Secular Rise in Sex Ratio. There has been a long-term increase in sex ratio since the 1920s in Britain and also in the United States. Like all of the preceding influences, this is quite small numerically, fractions of a percentage point, and very large numbers of subjects are required to demonstrate its existence. In addition, many other influences on sex ratio have been claimed, but they are on much weaker ground than those mentioned, and remain to be confirmed. For example, among several thousand Harvard men, the stocky men had significantly more sons that the lean ones.

There are, however, two very recent significant developments. In 1971 an Englishman named William James proposed a new theory of sex determination, as follows: Owing to the biomedical environment in the female reproductive tract early in the menstrual cycle, sperm containing the Y chromosome survive better than those containing the X chromosome. Males are therefore conceived preferentially in this early stage. The more frequently sexual intercourse occurs, the more likely is a male to be conceived. This theory would explain the findings of Bernstein and Renkonen, since intercourse is more frequent early in marriage than it is later on. James's theory would also explain the fact that nonidentical, or fraternal (dizygotic) twins are of the same sex more often than half the time. Theoretically, one would expect that half of all fraternal twins would be of the same sex, the other half of opposite sex. James's argument is that the time in the menstrual cycle when conception occurs influences sperm survival in the female reproductive tract and favors fertilization of two eggs by the same type of sperm.

The other recent development (1972) has been the discovery by Casperson and Lech, in Stockholm, that the Y-chromosome fluoresces distinctively when stained with a group of dyes containing the three-ringed acridine nucleus, and called flavines. This makes it possible to detect the Y chromosome in sperm or in tissue culture. We can finally settle the question of whether Shettles is correct. Shettles has reported two kinds of sperm, with round heads and narrow heads, and proposed that sperm shape reflects its content of X or Y chromosome. Unfortunately, the fluorescent dye kills all cells, so that it cannot be used to produce children of the desired sex, although it can be used to validate other methods for separating X- and Y-bearing sperm.

Sex Differences in Disease

As mentioned above, males have a higher mortality rate at all ages. This sex discrepancy is increasing in the United States, because death rates for women have fallen more rapidly than those for men (Table 11.1). Only in countries with low standards of obstetrical care have female rates exceeded male rates, and then only in the child-bearing ages.

This less fortunate male position is not the result of serious disadvantage in a few special cases of death; rather, mortality rates for males exceed those for females for most causes. Of the commonest causes of death, apart from those affecting the sex organs, only diabetes and hypertensive heart disease show higher death rates among females. Now paradoxically, females have greater morbidity—that is, they are sick more

Table 11.1. DEATH RATES PER 100,000
POPULATION FOR LEADING
CAUSES BY SEX IN 1973 [a]

	Male	Female
Heart disease	415.2	309.1
Cancer	187.1	148.4
Stroke	91.8	111.9
Accidents	78.7	32.8
Pneumonia	33.5	26.3
Homicide	15.5	4.3
Suicide	17.7	6.5
Diabetes	15.3	20.9
Cirrhosis of the liver	21.3	10.7
Emphysema	17.6	3.9

[a] SOURCE: *Statistical Abstracts of the U.S.*, 1975.

often with nonfatal or long-lasting chronic diseases (Table 11.2). They predominate in thyroid disease, diabetes, gallbladder disease, obesity, arthritis, psychoneurosis, and many benign tumors. Only a few illnesses are predominantly male in morbidity data, including peptic ulcer, hernia, accidents, and heart desease. Women get sick more often, but men do the dying.

Explanations of Disease Differences. The chromosomal basis of sex shows why the spectrum of disease distribution can have theoretical limits of 100 percent male and 100 percent female prevalence. (The word *theoretical* should be stressed, because few if any diseases other than those of the reproductive tract are limited exclusively to one sex.) As you know, there are twenty-three pairs of chromosomes in man, twenty-two pairs of autosomes, and one pair of sex chromosomes, the X and Y chromosomes.

Table 11.2. SOME CURRENTLY SUSPECTED
SEX DIFFERENCES IN DISEASE [a]

Predominantly Male	N	Ratio *Male : Female*
Infectious croup	850	2.5:1
Bacterial meningitis	201	1.6:1
Staphylococcus infections	1804	1.5–1.9:1
Respiratory distress syndrome (in newborns)	136	1.7:1
Kala-azar	120	2:1
Lymphoma	1000s	1.5:1
Cancer—liver	1000s	1.6:1
Cancer—pancreas	425	2.0:1
Cancer—brain tumor (length of survival)	100	1.7:1
Hyperactive children	750	6:1
Predominantly Female		
Subacute hepatitis	97	1.0:7
Pigment degeneration of the eye	202	1:1.6
Congenital dislocation of the hip (newborn)	64	1:4.5
Certain spinal abnormalities (Spina bifida)	3521	1:1.25
Depression	93	1:5.6
Cancer—thyroid	343	1:2.3
Cancer—gallbladder	100s	1:5
Cancer—lymphoma (length of survival)	100	1:2.5
Urinary tract infection (recovery)	133	1:4

[a] From various sources.

Males have the *XY* pattern, females the *XX*. Obviously, any genes peculiar to the *Y*-chromosome will occur only in males and will be transmitted by a father only to his sons. Some genes do occur on the *Y* chromosome, notably those favoring the slower maturation rate found among males. Such genes have a physiological effect. Hairy ears are said to follow a *Y*-linked form of inheritance. You also know that a behavioral abnormality has been found among some men with the *XYY* constitution. These men are supermales, with an extra *Y* chromosome. They are apt to be very tall, strong, often have acne, and in some cases are uncontrollably violent. Occasional senseless crimes of violence have been attributed to such individuals. However, no specific diseases have been assigned to the *Y* chromosome.

The commonest kind of sex-linked disease is of course the recessive located on the *X* chromosome. Because the male has only one *X* chromosome, he will show the disease if he receives only a single dose, on his *X* chromosome, from his mother. The classic example is hemophilia; another is red-green color blindness; there are many more. There are even some *X*-linked dominant diseases, which will not be discussed here.

Sex-Controlled Traits and Diseases. Most diseases occurring predominantly or almost exclusively in one sex are not so much sex-linked (occurring because of genes on the *X* chromosome) as they are sex-controlled. That is, the sex of a person controls, via his or her hormones, the expression of whatever genetic tendency may be present. Examples of sex-controlled normal traits include baldness, beard development, and pitch of the voice. Boys, eunuchs, and women usually lack beards and do not become bald. But given the appropriate hormone stimulus—testosterone, the male sex hormone—or following the menopause in the case of women, all three groups—boys, eunuchs, and women—can develop beards and baldness. These are sex-controlled normal traits. Sex-controlled diseases include gout and rheumatoid arthritis of the spine. In both of these, the proportion of men to women is about 19:1.

Predominantly Male Diseases. It is an intriguing fact that most of the inborn errors of metabolism which do show a sex difference occur mostly among males (Table 11.2). Many more male diseases might be included, notably Hodgkins' disease and the lymphomas. Perhaps the most important of all sex differences in disease, from an evolutionary point of view, has been discovered recently: the greater susceptibility of male infants and children to a variety of acute bacterial infections, particularly meningitis, tracheolaryngitis, and staphylococcal infections. Males are between 1.5 and 2 times more susceptible than females.

Pointing in the same direction is the recent discovery that females normally have more of the M-immunoglobulins than males. These substances in the blood protect against certain bacterial infections of the gastrointestinal tract, such as typhoid and paratyphoid fever, and shigellosis—bacillary dysentery. These findings suggest how, if not why, males have greater mortality than females.

Predominantly Female Diseases. Table 11.2 shows another remarkable fact which should also lead to fruitful hypotheses and investigation: the preponderance of autoimmune diseases among females. In these disorders, which have only recently been recognized, the body becomes sensitized to some of its own products, or antigens, reacting with the production of antibodies. An active research field at the moment is looking for autoimmune diseases in persons who are sexually anomalous because of their chromosomes, such as the Turner (XO) and Klinefelter (XXY) syndromes.

In most major disease entities, the genetic component, while it may be present, is not so clearcut, or so independent of environmental factors as it is in gout or hemophilia. Hypertension (high blood pressure), atherosclerosis, diabetes, and tuberculosis fall into this group of major diseases where the genetic component is probably multifactorial or polygenic. In addition, the environment plays a much greater role than it does in the single gene disorders. I have already stressed this vital point, which cannot be repeated too often: most of the interesting and important human characteristics are under polygenic control or influence. Such characteristics include intelligence, body size, shape, and functions, and the major diseases. All are polygenic, with strong environmental influence as well.

The explanation of many sex differences in disease prevalence remains unknown. Genetics can account for a small number of fairly rare conditions, as we have seen, and body constitution can account for some others. For example, women are more subject than are men to urinary tract infection, due to the short, relatively exposed female urethra. Women are more likely than men to fracture bones when elderly because women lose bone mineral after the menopause. But we are still missing many important clues in the sex distribution of disease.

Differences in Organ Susceptibility. In the entire gamut of thyroid disorders, benign and malignant, hyper- and hypofunction, infectious and noninfectious, women greatly predominate. This indicates an organ susceptibility associated with sex. The same is true for the gallbladder; for all its diseases females are again more susceptible. It is conceivable that the male preponderance in most lung diseases may be due to a similar sex-specific organ susceptibility as well as to greater smoking. Frequency, morbidity, and mortality are not the only characteristics of a disease; severity and course are also important. In general, women are said to do better with diabetes, hypertension, and rheumatoid arthritis than do men, although quantitative evidence is scanty.

Sex Differences in Physique and Physiology

Here one must think in terms of modes, trends, or ideals, not of absolute male and female types. While there is no doubt about the physical masculinity or femininity of most people, there is a gradient in both sexes away from or toward the opposite physique. Table 11.3 shows the range of some selected measurements among men and women. One looks at the ratio of hip and shoulder breadth, breast development, inner surfaces of thighs and calves, and relative length of trunk and limbs.

Typically, the sexes differ physically in three ways: (1) size, (2) shape, and (3) tissue structure. The most striking physical differences, the man's greater stature and broader shoulders, and the woman's broader hips, arise at adolescence; other differences appear earlier. Some, like the external genitals, arise before birth while others, like the relatively greater length of the male forearm, develop throughout the growth period.

James Tanner classifies the four mechanisms which bring about sex differences as follows: (1) Some differential growth rates operate only at adolescence, as a direct result of the differential sex hormone production at that time—androgens and estrogens. Hip and shoulder breadth fall into this group. (2) The male adolescent growth spurt occurs later than the female. The epiphyses of the long bones close and growth stops as a result of adult sex hormone production. Consequently, the longer period of male growth leads to greater size. In addition, the growth rates which operate equally in both sexes to change the infant's proportions to those of the adult continue to act for a longer time in the male. They include lengths of arms and legs relative to the trunk, and size of head relative to the rest of the body. (3) Some differential growth rates occur from birth or even earlier—for example, forearm relative to leg or whole arm length. (4) Finally, differential growth can occur at some particular time other than adolescence. For example, the penis develops at a particular phase of fetal life.

Let us examine each mechanism in turn. Mechanism (1) is caused by

Table 11.3. SEX DIFFERENCES IN PHYSIQUE:
SOME SELECTED BODY MEASUREMENTS [a]

THIGH CLEARANCE HEIGHT (SITTING) (INCHES)

	Percentile		
	5	50	95
3091 civilian males aged 18–79	4.3	5.7	6.9
3581 civilian females aged 18–79	4.1	5.4	6.9

SHOULDER BREADTH (INCHES)

1959 male railroad travelers, mean age 38	16.4	17.6	19.2
1908 female railroad travelers, mean age 35	14.4	15.7	17.6

HIP BREADTH (STANDING) (INCHES)

3328 Air Force basic trainees, mean age 18	12.1	13.3	15.0
850 WAF basic trainees, mean age 18	12.5	13.5	15.4

CHEST DEPTH (INCHES)

119 male Univ. of Chicago students	6.9	8.0	9.2
1013 female college students, midwest	6.4	7.3	8.2

[a] SOURCE: Damon, Albert, Stoudt, Howard, W., and McFarland, Ross A. 1966. *The human body in equipment design.* Cambridge, Mass.: Harvard University Press.

differential hormone secretion at adolescence, androgens versus estrogens, aided perhaps by differential response of tissues. Incidentally, we shall see later that differences in tissues or end-organ response to hormones accounts for some physical differences between populations as well as between the sexes.

Mechanism (2) results from slower maturation in the male. At birth and even before, girls are more advanced in skeletal maturation than boys and remain so up to adolescence. Every tooth appears earlier in girls, by periods which vary from two months for the first permanent molar to eleven months for the permanent canines. Thus, at any given chronological age, girls are further along toward maturity: the physiological age is greater. The male retardation is probably caused by genes on the Y chromosome, as we have seen.

The causes of mechanisms (3) and (4) are largely unknown. Genes on the Y chromosome cause the development of the penis by causing a fetal testis to develop and secrete a fetal androgen. The double dose of the X chromosome, among normal females (XX) and males with Klinefelter's syndrome (XXY) is thought to inhibit muscular development. So women are on the whole less muscular than men. The continually greater growth of the male forearm may be caused by a greater growth gradient in the local area. Women have reduced numbers of phalanges in the toes more often than men, and the second finger is usually longer than the fourth; in men, the opposite holds. James Tanner suggests that in women, more than in men, the available material seems to be exhausted before the distal or outermost part of the limb is complete.

With regard to growth in general, we saw earlier the catch-up growth following malnutrition, injury, or disease once the disturbing factor has been removed. Girls are harder to deflect from their genetic growth pattern than boys; their growth is less affected by environmental stimuli, whether good or bad. Boys show a greater response to malnutrition and a faster and greater recovery.

Differences in Physique

Body Form. Size has already been mentioned as one of the differences between the sexes. To nobody's surprise, there are also differences in shape; in fact, the clothing and advertising industries are based on these differences. In the skeleton, men have broader shoulders, narrower hips, longer legs, still longer arms, and most particularly longer forearms. In women, the increased hip breadth is due mostly to pelvic bones, which enlarge at adolescence due to estrogen secretion. The pelvis of the hypogonadal (Klinefelter or XXY) male is like that of a normal male. That is, estrogens are needed, and not just the double X chromosome genetic constitution, for a broad female type of pelvis. The male skeleton is denser than the female, due to the thicker cortex, or solid part of each bone. This difference in bone density arises at adolescence.

Soft Tissue. Most of the sex differences in amount and distribution of soft tissue arise at adolescence, but not all. Girls have somewhat more subcutaneous fat before adolescence, and much more after it. Calf and forearm X rays show that, from early childhood, boys have slightly more bone and muscle. At adolescence, males increase their bone and muscle and *lose* fat, whereas females increase very little in bone and muscle and *gain* fat. Thus it is possible to judge sex correctly from limb X rays only 60 percent of the time in seven-year-old children—little better than pure chance—whereas 95 percent of adult X rays can be correctly sexed, by inspection.

In the young American male 15 percent of body weight is fat; in the young female, 30 percent. These values of course increase for both sexes with age. The differential development of fat at adolescence—males losing fat, females gaining fat, accounts for part of a characteristic sex difference in adult shape, first pointed out by William H. Sheldon. But there are skeletal shape differences as well. When the heels or ankles are placed together, the thighs in the typical female touch, and only a small space appears between the knees and ankles. In men, there is space between the thighs and much more between the calves. Serial photographs (i.e., of the same person) show that this difference is due mainly to skeletal structure, rather than to fat. An individual's characteristic pattern is developed in its essentials long before adolescence. So the differential development of fat at adolescence accounts for only part of this difference in space between the thighs.

Before puberty, there is little difference between the sexes in the shoulder muscles, but boys have stronger hand grips than girls. This doubtless results from their longer and more muscular forearms. After puberty, of course, men on the average are much stronger than women. Boys are stronger than girls in all cultures, even in simple societies where girls do more work. Many other physiological and biochemical differences have been reported between normal men and women (Table 11.4).

Nerve conduction velocity is said to be slightly faster in women than in men: 54.7 meters per second for women, 51.2 for men. This is a greater effect than that for age, recorded over a range of thirty to ninety years for men. Men have higher amino acid levels in the blood, more calcification of the aorta and coronary arteries. Men have greater work capacity (that is, they have higher oxygen uptake and lower pulse at maximum work load). Men can adapt better to heat—that is, they produce more sweat at lower temperatures although the two sexes have the same number of sweat glands. In addition to differences noted above (growth hormone, M-immunoglobulin, late thinning of bone), women have more ascorbic acid (vitamin C) with equal intake.

Table 11.4. SEX DIFFERENCES IN SOME
PHYSIOLOGICAL AND
EPIDEMIOLOGIC CHARACTERISTICS

Trait	Male Proportion or Ratio
Blood	
Amino acid	Higher
Growth hormones	Resting—equal
	Ambulatory—lower
Immunoglobulins	IgG, IgA—equal
	IgM—lower
Palmar sweat	Lower
Costal cartilage calcification	Lower (1:5—ages 30–50)
Aortic and coronary calcification	Higher
Femoral fractures	Below age 40, higher;
	above age 40, lower
Fetal and infant mortality	Higher
Serum lactic dehydrogenase	
alpha$_1$:alpha$_2$	1:1.4
Ascorbic acid (same intake)	
(ages 13+)	1:2

Behavior. For a final item, it has been shown that in many mammals, including rodents and monkeys, not only do the sexes differ in structural details of the brain and nervous system, but the behavior patterns determined by the brain are basically female. Male patterns are induced by the action of the male sex hormone, testosterone, on the brain of the newborn animal, or even earlier. However, testosterone supplied after a critical period of brain development will not produce male behavior. This raises the possibility that sex behavior, including homosexuality, may depend to some extent on the hormonal makeup of the individual during the development of the nervous system in the fetal or new-born period, and not only in later stages.

So the sexes differ profoundly, on a genetic basis, in many aspects of biology and behavior. The human infant is not a blank tablet, waiting to be stamped male or female by the social environment. Long before birth and throughout subsequent life, a person looks and acts in accordance with his genes, and hormones. Our biological inheritance sets limits to our interaction with the environment.

SUGGESTED READINGS

Harrison, G. A., Weiner, J. S., Tanner, J. M., and Barnicot, N. A. 1964. *Human biology*. New York: Oxford University Press.

Money, John. 1975. *Sexual signatures, on being a man or a woman*. Boston: Little, Brown.

Money, John, and Ehrhardt, A. A. 1972. *Man and woman, boy and girl*. Baltimore: Johns Hopkins University Press.

Stern, Curt. 1973. *Principles of human genetics*, 3d ed. San Francisco: Freeman.

12
RACE, PHYSIQUE, AND DISEASE

A third major axis of human variation is race. Here we are concerned with differences among human groups; chiefly physical differences and differences in disease. I shall merely mention in passing, for lack of space, the equally important differences in physiology and biochemistry, subjects covered at length in books of other kinds, especially those on human genetics. To begin, we must face up to some matters of definition. Everybody knows what age is. Definitions of infancy, childhood, adolescence, maturity, and old age can be reasonably established and calmly discussed. Definitions of male and female are also fairly well agreed upon, although they may be less absolute than used to be thought. With race, however, we enter an area where real differences of scientific opinion exist among human biologists. In addition, the concept of race is heavily charged with emotion, for reasons outside biology. We are all aware of what those reasons are. But whatever our interest in race, objective examination of the facts of human biological variation, and of the mechanisms bringing about this variation, cannot fail to be useful.

The Nature of Race

Reasons for Study. First of all, why does a human biologist study race? For four reasons. The first is purely descriptive, or taxonomic; the biologist wishes to catalog and classify the variety of natural human populations, in some orderly fashion. As William S. Laughlin has put it, we wish to document the human genetic repertoire. If men were birds, nobody would question the purpose or value of taxonomic description.

The second reason for studying race is evolutionary, or phylogenetic, or historical. Here one tries to determine or to reconstruct the relationships of the various human populations to one another, both in time and in space. Did population A develop from population B, or did both split off from a common ancestor, C? Did this occur at the same time, or did population A branch off earlier than B?

The third reason concerns the dynamics, the mechanisms, and the processes underlying the observed variations. How did the present differences among the populations of man arise? Such mechanisms include the familiar biological ones of evolution and adaptation: namely, mutation, selection, migration, and genetic drift, as well as the human mechanisms of culture, and of social and sexual selection.

And finally, the fourth reason for studying race is that of practical helpfulness. We know that breeds of dogs—bloodhounds, pointers, beagles—(which are artificially produced races) have distinctive characteristics and abilities. If we see a dog that looks like a pointer our best guess is that it can be trained to stalk sitting birds better than a fox terrier. Now it would be outrageous to suggest that races of man are so distinctive in their traits, that each individual assignable to a race will have a given set of traits or abilities. Good long-distance runners are lanky and long-legged regardless of skin color, and great singers depend on vocal equipment, not race. Nonetheless some special qualities of race have come to light. For reasons not known, aboriginal Australians are found to have a significantly faster water turnover than white Australians no matter how acclimated, and they can also inhibit sweating in a water-saturated environment better. So different races may serve as presorted groups having a higher probability of exhibiting certain traits, if investigation shows such a probability. Thus, practically, we may try to use observed and established variations among human groups (not necessarily by race alone) to design equipment; to select persons for the occupations or the geographic areas to which they are physically and physiologically best suited; and, with regard to disease, to detect, prevent, diagnose, and treat. All these valid reasons for studying race mean that the human biologist or physical anthropologist is not being a frivolous stamp-collector or a reactionary racist when he describes and tries to explain racial variation in man.

Definitions of Race. Although modern definitions of race are numerous, they all include two facets: namely, (1) membership in a breeding population with (2) distinctive biological characteristics. This was the definition of R. B. Dixon sixty years ago, E. A. Hooton forty years ago, and the definition of Ernst Mayr, Theodosius Dobzhansky, and others today. The main difference is that nowadays the emphasis has shifted from describing the biological characteristics to delineating the breeding structure of the populations involved. Even descriptively, we know a great deal now about biological characteristics other than the metrical and morphological ones that were the sole concern of early anthropologists. Different proportions or genes in blood groups and serum proteins, for example, are now used to characterize different human groups, in addition to skin color and hair form. The recognized major races are by no means homogeneous in many of these traits. We are also much more interested in the dynamics of race formation than in pure description, but our knowledge of mechanisms lags far behind our interest. This is an exciting field for research, but one which will not be possible to study under natural conditions very much longer. We shall have to move into the laboratory. Selection and evolution are still occurring in man, to be sure, as we have seen in earlier chapters. But as civilization spreads, environments are becoming more uniform. The mechanisms of natural selection and race formation are less and less affected by the natural habitat and climate, and more and more by the man-made environment.

To illustrate the speed of civilization's spread, let me mention the recent experience in the Solomon Islands of a Harvard graduate student of social anthropology. His professor advised him to study a particular group, the Nasioi on Bougainville Island, who were alleged to be still living in unspoiled aboriginal simplicity. When he arrived to take up residence, he was welcomed by the village head man, who opened his refrigerator and offered him a can of cold beer! This kind of experience lends a sense of urgency to the study of those few remaining groups that still live as man has lived for most of human existence, in local, largely self-sufficient communities. The point is that fewer of the evolutionary mechanisms will remain to act on race formation. Mutation and migration, or gene flow, will remain, but isolation, genetic drift, and the various forms of adaptation to the natural environment, physical and biological, will no longer occur or will be much diminished (Figure 12.1). To exaggerate only a little, the Solomon Islanders will live in air-conditioned, malaria-free houses and walk around wearing clothes like ours, while the Eskimo will have both refrigerators and central heating. All will receive immunizations, penicillin, and vitamin supplements when young and doubtless other doses when they grow old. It is therefore unlikely that we shall have a chance to observe any of the stages in the differentiation of pigmenta-

12.1   Some of the racial elements in South America—Indian, black African, Chinese. The girl at the right is the daughter of an Indian mother and a Chinese father. (courtesy Peabody Museum, Harvard University)

tion, hair form, facial features, and biochemical adaptations that distinguish races today. And even more to the point, it is questionable whether we shall ever be able to explain some of them.

Kinds of Differences. How do races vary? They vary in physical appearance; in characteristics of the blood, such as the various blood groups and serum proteins; in physiology; and in biochemical characteristics, such as ability to taste certain substances, or the amino acids excreted in the urine, or adult intolerance to lactose, the sugar found in milk (Table 12.1). (Orientals have low tolerance; the question arises whether Vietnamese adopted children should be given quantities of cow's milk like our own offspring.) Races differ in a myriad of other genetic details, including susceptibility to disease and response to disease.

Races are by no means all-or-none groupings in any of these respects, which is a probable public stereotype or misconception. Races are not types, pure or impure; they merely differ in the frequency of various particular genes, doubtless a small minority of the whole genotype. Sometimes there is virtually no overlap between races, as in pigment and hair form between unmixed black Africans and Northwest Europeans. A good

Table 12.1. SOME RACIAL DIFFERENCES
IN PHYSIOLOGY
AND BIOCHEMISTRY

	In Relation to Caucasoid Norms	
	Negroids	Mongoloids (Orientals)
Birth: weight	−	−
Birth: skeletal maturation	+	
Birth: dental maturation	+	
Birth: neurological maturation	+	
Neonatal motor development	+	
Acuity, auditory	+	
Acuity, visual	+	
Blood pressure	+	−
Bone density	+	−
Color blindness	−	−
Keloid formation	+	
Lactase deficiency (adult)	+	+
Pulmonary function	−	
Serum globulins	+	
Skin resistance, electrical	+	
Tasting, phenylthiocarbamide	+	+
Twinning, dizygotic	+	−

deal more often, there is overlapping, quite separate from anything due to mixture. Differences between racial averages are in most features modest in magnitude even when statistically significant. Differences in many of the blood group distributions are greater between local populations than between the racial groups they make up.

At this point, a word is in order to recall the understanding of genotype, or genetic composition, versus phenotype, or the manifestation of the genes. Sometimes genotype and phenotype coincide, as in the case of homozygous recessives, like albinism or blood group O. Sometimes genotype and phenotype do not correspond, as in the case of blood group A, which may have the genotype AA or AO. The dark skins of Africans and of some Melanesians are probably due to different sets of underlying genes. Now the classical physical descriptions of race are based on phenotype, while much of our interest in race, but by no means all, concerns the genotype. Tracing relationships of the various races of man to one another requires knowledge of the genes. As we have just seen, the phenotype is useful in providing clues to the genotype. But the phenotype is even more important than that because natural selection works on the phenotype. Disease works on the individual's phenotype, not on his genes. As already mentioned, a person may have a truly superior set of genes for intelligence, but they will do him and his population group no good whatsoever if they are associated with, for example, albinism or diabetes in any technologically primitive group. Such a person is likely to die young, or to be forbidden to marry, and his other, possibly superior genes will be lost. So phenotype is not merely skin deep, any more than race is. What we see, or the way in which the genes are expressed, is vital in the study of human variation, whether that expression is manifest as body measurements, as physical features, as biochemical or physiological differences, or as differences in disease susceptibility and resistance.

The scientific controversy surrounding the definition of race has two sources: (1) The various traits used to distinguish breeding populations do not always yield the same groupings (Figure 12.2). That is, two populations can be similar in the *ABO* blood groups, but very different in others, such as the Rh system. (2) Most of the physical features used to describe races are polygenic. The genes which determine these features are multiple, and of uncertain number and mode of inheritance. Furthermore, these same features, such as skin color, height, and head shape, may be influenced by the environment. Both the polygenic inheritance and the environmental influence of physical features are in marked contrast to simple and well-understood single-gene systems like the blood groups and serum proteins. For these reasons, certain anthropologists would throw out the phenotype, all the traditional polygenes, and the whole concept of race, in favor of using single-gene traits only. However, a few efforts to arrange human populations in this way have failed—as, for example, William C. Boyd's 1950 attempt to classify races by blood group alone (see Figure 12.2). Boyd has admitted the failure, but some writers still include Boyd's classification. This atomistic approach would give a different racial picture for each blood group system and for each of the many other separate traits that distinguish breeding populations. In fact, the notion of human variation, that is, differences among people—or differences among breeding populations—would become a study of clines, which are geographic changes in frequency of separate genes. This would produce a hopelessly confused picture, losing even the rough correspondence between geography and human biological constitution that the present classification delineates. Genes do not exist in isolation; they occur in a matrix of other genes. The whole genotype is carried by—although incompletely manifested in—the phenotype. By taking single genes or even single traits, out of context, you become a poor geneticist, as well as no anthropologist at all.

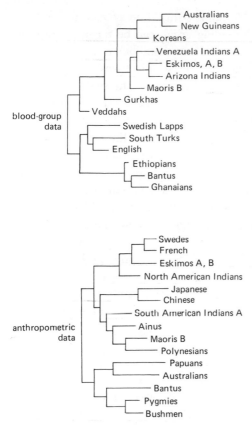

12.2 Computer-based reconstructions of human phylogenies. The first uses data on gene frequencies at five blood-group loci and the second a set of some twenty-six anthropomorphic measurements including stature, weight, skeletal dimensions, skin color, and hair thickness. Statistical estimates were made of the degree of resemblance among populations studied. Note the equally startling similarities and differences. (From Lerner, I. Michael. 1968. *Heredity, evolution and society.* San Francisco: Freeman, p. 230, using data of L. L. Cavalli-Sforza and A. W. F. Edwards.)

To be sure, there are many definitions of race. They are arbitrary. Most races do overlap in the distribution of most traits. There are some groups intermediate between the classical races, and some groups are difficult to classify. But all of these objections can be met. The arguments have really been about systems of classification, not about the existence of physical and physiological differences, nonrandomly distributed, among groups of men. All definitions and classifications, of any kind, are arbitrary; the important point is whether they are useful. To take a simple example, it is impossible to set a single height at which shortness ends and tallness begins. But this does not negate the existence of short and tall men, or the concept of tallness. The same holds for the racial groupings of mankind. Of course all intergrades exist, because races are statistical or populational concepts, not typological concepts. It is incorrect to consider races as fixed or once pure types. All human populations, even inbred isolates, are genetically variable. But even though distributions of some traits may overlap between races and populations, the averages may differ significantly.

One final word regarding definitions, this time of race and ethnic group. A biologically distinct or breeding group can be termed a race or population; such a group has a relatively large number of genes in common, and usually has some distinguishing physical features as well. A culturally distinct group is an ethnic group. Frequently the two kinds of homogeneity, biological and cultural, overlap or coincide, as in the case of an American Indian or African tribe, or even a linguistic, religious, or colored minority in a city or country, provided only that the members of the group share a common ancestry. The Boston Irish, Italian, Jewish, Armenian, Chinese, or black communities, for example, are varyingly distinctive both biologically and culturally, in widely varying degrees—they are racial as well as ethnic groupings. Certainly no one could separate all persons of Irish and Italian descent at sight, but an adult Bostonian, especially if a member of either group, might do so fairly well.

Biological and cultural similarity can reinforce one another. On the one hand, distinctive physical features make it easy for society to practice or enforce certain breeding patterns; for example, intermarriage or interbreeding between whites and nonwhites is now forbidden in South Africa. On the other hand, distinctive cultural practices, like religion, may isolate inbreeding populations more or less completely, as in the case of the Jews for many centuries, or small contemporary groups like the Amish, Dunkers, Hutterites, and Mennonites. Even such an apparently minor characteristic as accent seems to be an effective bar to panmixia (complete interbreeding) in Britain. Whatever causes or maintains a group in relative isolation, social or biological, the net result is a group that has more genes or social practices, or both, in common than has the population at large.

This situation is not difficult to understand, but it has been confused by a recent tendency to substitute the term *ethnic group* for *race*. This effort reflects concern with the evils of race prejudice. Ashley Montagu has been one of the leading proponents of this point of view. It is doubtful, however, that abolishing the word *race* will abolish race prejudice. If a new term is desirable, *ancestral group* or *population group* or *breeding group* or even *biological group* would retain the important notion of biological distinctiveness, which *ethnic group* does not.

Race and Physique

Everyone agrees that we can recognize at least three major racial stocks: the Caucasoids, the Mongoloids, and the Negroids or, familiarly, whites, yellows, and blacks. Generally writers make inclusions in these: the people of India in the Caucasoids in spite of dark skins, and the American Indians in the Mongoloids in spite of some other differences. But including Melanesians and aboriginal Australians with Africans as Negroids is a dubious matter. In any case, discussions of race have tended to founder on attempts to classify instead of simply noting that human populations vary over the map of the globe. Classification is quite unnecessary and in fact undesirable for our purposes here, and an attempt to delineate the peoples of the world is more than we can undertake. Because of good magazines and good television documentaries, most readers surely have a good idea of the outward appearance of many different local races or populations.

Racial Features. For recognizing differences of populations, skin color and hair form have always been the most widely useful features, followed by head shape, nose shape, and the amount of body hair in males. Some special features are found in one or another racial group, notably the Mongoloid eyefold and their pigment spot, on the lower back in infancy, the small stature of some pygmy groups in Africa and Asia, or the steatopygia or fatty buttocks of some relict populations in Africa (Bushmen and Hottentots) and in the Andaman Islands (Figure 12.3). In addition, we have already seen that the melanosomes, or pigment granules within the skin cells, are packaged singly in Africans and Australian aborigines, but in multiples of two, three, and four in other groups.

a   b

12.3 Steatopygia, or the deposition of fat in the buttocks, is commonest among South African Bushmen, Hottentots, and Andamen Islanders of the Indian Ocean. It is tempting to speculate that steatopygia has survived from very early times, since women exhibiting the trait resemble so-called fertility figurines from prehistoric times, but this cannot be proven.

Mechanisms Selecting Physical Features. Four main mechanisms might explain certain of the physical features we have been discussing. First, sexual selection may be responsible for man's unique loss of body hair among the primates. This is what Darwin believed, since it is hard to see what advantage man has gained by losing hair. Breast development in human females is probably another feature resulting from sexual selection. In the great apes and other primates, the females, even when lactating, or suckling their young, have breasts little or no larger than those of the males. Anatomically, the human breast is mostly fat, with very little functional glandular tissue. So we might conclude that the functions of hairlessness, steatopygia, and the human breast are mainly ornamental.

Another mechanism explaining certain distinctive traits is the failure of end-organs to respond to adequate amounts of hormone. The pygmy is small not because he lacks growth hormone; he has plenty, but his long bones do not respond to it. Administration of human growth hormone does not make pygmies grow taller. American Indians for the most part lack beards. They have plenty of androgen, the male sex hormone which promotes beard formation, but their hair follicles do not respond to it.

The third and most important mechanism is adaptation to the environment. We shall see later on how skin color represents an adaptation to sunlight, nose form to humidity, and body build to temperature.

The fourth and final mechanism underlying physical features is a collection of unknowns. We have no really satisfactory explanations for population differences in such diverse features as hair form, lip thickness, tooth size, jaw protrusion, or ear form. Perhaps the most plausible explanation is pleiotropy—that is, they represent secondary effects of genes whose primary effects, which we have not yet discovered, are advantageous.

Race and Disease

Associations of race and disease can be mediated by a variety of mechanisms, as we shall see: anatomical, biochemical, and genetic. The recognition and study of the racial associations of disease can help the practicing physician in diagnosis, prognosis, and treatment. Knowledge about sickling crises and Mediterranean fever can prevent unnecessary surgery. And we have seen that fever, abdominal pain, and an increase in the white blood cells may have different causes in a North European, a Mediterranean, or a person of African descent. In a North European, this combination would suggest a "surgical" condition such as appendicitis or an inflamed gallbladder. In a Mediterranean, it might also represent a hemolytic crisis due to deficiency of the enzyme glucose-6-phosphate dehydrogenase (G-6-PD), or one of the periodic crises of familial Mediterranean fever, a condition virtually confined to Syrians, Armenians, and other groups in the Eastern Mediterranean. In an American or African black, the same combination of symptoms could signify appendicitis, gallbladder inflammation, or a sickle-cell crisis, in which red blood cells having the sickle form of hemoglobin are destroyed in large numbers. A physician who is aware of racial and ethnic patterns of disease can thus spare his patients from an unnecessary operation.

Knowledge that certain diseases are concentrated in certain groups within a nation or region can help the health administrator concentrate his case-finding, treatment, and preventive facilities among those groups that need the most attention. With finite resources, it makes great economic and humanitarian sense to conduct campaigns to detect alcoholism, dental caries, cancer of the uterus, tuberculosis, or venereal disease among the segments of the population where these are most likely to be frequent. Information on the racial or ethnic associations of disease can help not only the practicing physician and the health administrator, but also the epidemiologist and other medical scientists to determine the cause of the disease in question.

Racial Associations with Disease. For a first general picture we might note that in the United States, overall mortality rates are greater for nonwhites than for whites, at virtually all ages (Figures 12.4 and 12.5). U.S. nonwhites are 92 percent blacks of African descent, with a scattering of American Indians and Orientals.

Racial associations with mortality are obvious. Due to the greater reduction in nonwhite mortality between the two periods 1940 and 1971, the nonwhite rate was closer to the white rate in 1971 than in 1940, but the nonwhite mortality was still much greater. The explanation is largely or wholly environmental.

When we examine *mortality* data by specific causes of death, some interesting patterns appear (Table 12.2).
Notice: (a) the higher rates of several infectious diseases in nonwhites, for instance, pneumonia, tuberculosis, syphilis; (b) high rates of arteriosclerotic heart disease in whites versus high rates for hypertension and cerebrovascular strokes in nonwhites; and (c) differences in cancer of various sites, particularly leukemia, urinary-tract cancer, and genital-tract cancer.

Religion, Ethnic Group, and Race.. Ethnic background and religious

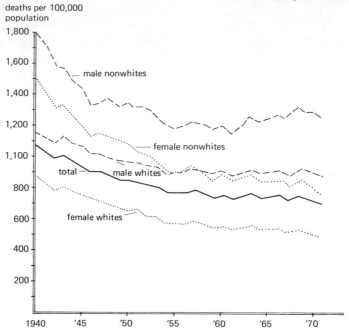

12.4 Death rates, 1940–1971, by sex and race. (From *Social Indicators*, a publication of the Social and Economic Statistics Administration, U.S. Department of Commerce, 1973. Note that the U.S. Census Bureau gives figures separately for "white" and "nonwhite" groups.)

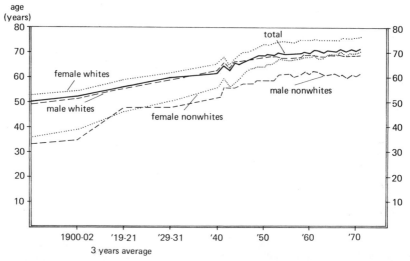

12.5 Life expectancy at birth, 1900–1971 by sex and race in the U.S. (From *Social Indicators*, a publication of the Social and Economic Statistics Administration, U.S. Department of Commerce, 1973. Note that "nonwhite females" have now surpassed "white males," and that "nonwhite males" have reached the "white male" level of about 1940.)

Table 12.2. DEATH RATES FROM SELECTED CAUSES
FOR WHITES AND NONWHITES
IN THE UNITED STATES, 1965
(PER 100,000 POPULATION
IN SPECIFIED GROUP)

Cause	Mortality Rates White	Mortality Rates Nonwhite	Rate in Whites as Percentage of Rate in Nonwhites
Suicide	11.9	5.0	238
Leukemia	7.4	4.1	180
Arteriosclerotic heart disease	303.8	175.8	173
Urinary cancer	7.5	4.6	163
Lymphosarcoma	8.1	5.1	159
Breast cancer	14.6	9.5	154
Peptic ulcer	5.6	3.8	147
Respiratory cancer	27.7	20.6	134
Digestive cancer	49.6	41.8	119
Motor vehicle accidents	25.3	25.8	98
Vascular lesions of the central nervous system	102.3	114.7	89
Genital cancer	20.5	23.3	88
Cirrhosis of liver	12.5	14.6	86
Diabetes mellitus	16.7	20.1	83
Accidents, other than motor vehicle	28.9	41.2	70
Pneumonia	28.9	44.2	65
Hypertensive heart disease	24.7	55.3	45
Tuberculosis	3.4	9.3	37
Syphilis	1.0	2.8	36
Homicide	3.0	24.6	12

belief are so strongly correlated that religion may sometimes be used as an indicator of ethnic groups. Ethnic groups are those with a common culture—they are not races. In most American communities, Catholic and Protestant groups include such a variety of ethnic backgrounds that these religions, by themselves, may be of little practical use in identifying groups that are ethnically homogeneous. On the other hand, the Jewish religion does identify a group which is more homogeneous than the population at large. This homogeneity has three components. First a common ethnic background, since most American Jews are Eastern European in ancestry. Second, religion or ethnic status may influence everyday practices to a greater extent than prevails for the general population. Examples that come to mind are circumcision, moderate use of alcohol and tobacco, and choice of occupation. Third, the Jews are also somewhat distinctive and more homogeneous genetically than the general population. This is apparent from their *ABO* blood distribution (specifically, being high in blood group *B*, which is characteristic of Asia), and also from their virtual monopoly of certain diseases known to be genetically determined, as we shall see shortly.

A great deal of interest in religious differences has arisen in the cancer field, probably because of the well-known rarity of cancer of the cervix uteri among Jewish women. This infrequency has been noted in many different countries and economic groups. An examination of mortality rates from cancer in New York City (Table 12.3) indicates that cervical

Table 12.3. ESTIMATED ANNUAL DEATH RATES
(PER 100,000 POPULATION
AGED FORTY-FIVE YEARS AND OVER)
FROM CANCER, NEW YORK CITY, 1955

Site	MALES				FEMALES		
	Catholic	Jewish	Protestant		Catholic	Jewish	Protestant
Cervix uteri					21	8	22
Tongue	9	2	8				
Mouth, pharynx	15	4	12	Jews			
Esophagus	21	13	21	less	3	6	5
Larynx	18	7	17				
Lung	50	34	56		8	8	7
Stomach	62	76	76		41	48	40
Rectum	42	40	45		30	30	27
Prostate	42	30	50	No			
Bladder	24	24	31	difference	10	7	14
Ovary					22	37	28
Lymphosarcoma	5	15	7		4	9	4
Leukemia	18	20	13		9	19	18
Glioma	4	7	5		4	6	4
Melanoma	2	6	2	Jews			
Kidney	11	18	15	more	5	8	4
Pancreas	28	38	28		19	26	19
Large intestine	56	84	57		59	77	59
All	585	595	638		437	511	468

cancer is not the only form of cancer to show striking differences, according to religion.

The low mortality from cervical cancer among women is clear. Low rates for cancer of the upper digestive and respiratory systems among Jewish males doubtless reflects their low intake of alcohol and tobacco, both of which are causative factors in such cancers. No such simple explanation is apparent, however, for the Jewish preponderance in various other forms of cancer, as shown in the bottom portion of Table 12.3. In another form of cancer too rare to appear here, cancer of the penis, Jews have much less than members of other religions, doubtless because of circumcision. Moslems, who are circumcised at ages from eleven to thirteen, have rates for cancer of the penis intermediate between those of Jews, or others circumcised at birth, and Hindus, who are uncircumcised.

Interpretation of Racial or Ethnic Associations. Before proceeding with even more striking differences in the racial and ethnic prevalence of disease, let us pause to consider the interpretation of such differences. To help interpret racial or ethnic associations with disease MacMahon, Pugh, and Ipsen have devised a check list of possible explanations. These are: (1) errors of measurement, (2) differences between groups with respect to more directly associated demographic variables, (3) differences in environment, (4) differences in bodily constitution, and (5) differences in genetic constitution. Let us review these in depth.

Errors of Measurement. The errors could also explain some reported age and sex differences in disease. With respect to age, old people are less likely than young ones to receive accurate diagnoses, and even age itself is less accurate. As for sex differences in disease, some harmless metabolic conditions have been falsely reported as more common among males, merely because males are more often examined than females, in connection with employment, insurance, or military service.

Racial or ethnic distributions of disease are even more likely to be distorted, due to inadequate diagnoses, differential access to medical care and utilization of medical care by racial or ethnic groups, and lack of precision in estimating the populations at risk. U.S. black persons appear, falsely, to have less leukemia and fewer congenital anomalies than white persons merely because they have had less ready access to diagnostic facilities. Even where facilities are equally available to all groups in the population, there may be differences in utilization. For example, in the same medical care plans in the United States, utilization rates tend to be lower for blacks than for whites, and lower for Catholics and Protestants than for Jews.

Differences Between Groups with Respect to More Directly As-

sociated Variables. Some of the reported differences between whites and nonwhites may reflect differences in age, sex, or socioeconomic status between the groups rather than differential disease susceptibility. For example, a group with a preponderance of children will have more ear infections than an older group. The older group will have more cancer and heart disease. A difference in disease rates between whites and blacks of the same age and sex should be pursued by comparing whites and blacks within similar socioeconomic or occupational groups. If the differences in disease rates persist, then it is unlikely that they are due to socioeconomic or occupational differences.

Difference in Environment. In accounting for racial or ethnic associations with disease, we want to know whether associations result from the environments in which the different groups are living at the time of observation, or whether they are a more permanent feature of the group, racial or ethnic, living in different environments—that is, we can look at migrant populations. For example, stomach cancer rates are much higher in Japan than in the United States. This could have either a genetic or an environmental explanation. Rates of stomach cancer are indeed higher for Japanese living in the United States than for American whites. But rates for Japanese living in Hawaii are higher than for Japanese living in the mainland United States, while rates for Japanese living in Japan are highest of all. Among the Japanese living in Hawaii, rates of stomach cancer are higher for those born in Japan than for those born in Hawaii. This indicates that the causal factor is associated with a way of life that is relatively susceptible to change on migration and is certainly not as constant as one would expect of a genetically determined trait.

On the other hand, there is some evidence that Japanese in America as well as in Japan are more subject to strokes than are white Americans. If the excess is constant among different groups of Japanese, wherever they live, one might then suspect a genetic basis. This seems to be the case with high blood pressure in West African blacks and their descendants in the Americas. In a wide variety of environments, such as North America, Panama, the Caribbean and in the whole spectrum of social stress from inner-city slums to quiet backwaters of idyllic Caribbean islands, blood pressure and strokes are higher among blacks than among whites. This constancy in a wide variety of settings suggests a genetic basis.

Environmentally caused differences between racial or ethnic groups can be shown by the large number of diseases related to socioeconomic status and to occupation. Most infectious diseases are more common to the poor, but several important causes of death are higher in upper socioeconomic classes. These include poliomyelitis, coronary heart disease, breast cancer, and leukemia. In relation to infection, perhaps the most obvious concomitants of socioeconomic status are crowding and nutrition. Inadequate nutrition causes vitamin deficiency diseases, like pellagra, as well as delayed and stunted growth. It also weakens resistance to any infection. Superabundant nutrition, especially when a high fat intake is combined with sedentary life, contributes to obesity, diabetes, and coronary heart disease. Diet undoubtedly contributes to ethnic variation in stomach cancer since high rates are found in Japan, Iceland, and Finland: a dietary component common to all three countries is smoked fish, which has been suggested as a contributing factor. Incidentally, the frequency of stomach cancer has been falling steadily in Britain and the United States for the past fifty years. This is obviously an environmental effect; genetic effects cannot occur so fast. Probably some change has occurred in our diet, perhaps less deep-fat frying of food.

Ethnic differences in personal customs and habits produce different environmental disease patterns. Alcohol and tobacco consumption are prominent examples; others are age at marriage, type of contraception, number of children, and infant feeding practices, breast versus bottle. These customs differ from one group to another and doubtless underlie many ethnic differences in disease. It has been proposed, for example, that the high rate of leukemia among Jews may result in part from overexposure to X rays in doctors' and dentists' offices.

One type of physiological characteristic having an environmental component which varies among racial, or more particularly among ethnic groups, is the level of immunity against microorganisms. For example, the low frequency of paralytic poliomyelitis in American Indians and blacks is attributed to frequent early exposure to the virus (as a result of overcrowding, unsanitary surroundings, and other features of low socioeconomic status) at a very early age when paralysis is rare. This early exposure results in subsequent high levels of immunity.

Differences in Body Constitution. This brings us closer to some biological associations of interest to physical anthropologists. We have already mentioned, in relation to sex differences in disease, that the shorter and more exposed female urethra results in more infections of the genitourinary tract among women than among men. Another example of a sex difference in disease resulting from differences in body constitution is the more frequent arthritis of the hip and pelvic joints among women. These are attributed to skeletal differences which result in different lines of stress imposed by weight-bearing. With regard to racial differences, the protection against the sun's rays afforded by skin coloring accounts for

Table 12.4. CERTAIN BONE DISORDERS
AMONG U.S. BLACKS

Disease	Age and Sex Affected	Black:White Ratio
Congenital dislocation of hip	Infant females	1:50
Legg-Perthes	Juvenile males	1:50
Traumatic fracture	Elderly females	1:50
Osteoporosis	Elderly females	1:5

much lower rates of skin cancer among darkly pigmented groups. The increased density of the bones of black persons may account for their remarkably low frequency of certain bone disorders, namely osteoporosis and traumatic fracture (Table 12.4).

Another possible example of diseases associated with differences in body constitution is upper respiratory infections or colds. Eskimos, with long narrow noses, are said to have frequent upper respiratory infections and chronic ear infections, whereas Africans and Malayans, with short, broad noses, are reputed to show these conditions only rarely.

Physiologically and biochemically, as we have said, there are marked racial differences. Some of these differences are associated with disease—for example, blood group O is associated with peptic ulcer and blood group A with cancer of the stomach. To the extent that such biochemical differences relate to disease they illustrate what is meant by "differences in body constitution."

Differences in Genetic Constitution. The last point on the checklist, differences in genetic constitution, of course determines many of the more directly associated attributes just mentioned, like skin color or bone density. In addition, ethnic or racial concentrations of genetically determined disorders are usually due to inbreeding, or the limitation of mating within the particular group over many generations. Genes which originally arose by mutation or which were introduced into the group will tend, as a result of inbreeding, to remain within the gene pool of the group. With single genes that are clearly expressed, the ethnic or racial concentration is apparent, and genetic analysis within the group, such as by twin and family studies, shows their genetic basis.

Differences in Prevalence and Severity. We turn now to more actual data on the racial and ethnic distribution of disease. As with disease distribution by sex, there is a continuous gradation from diseases that occur about equally in all groups to those found almost exclusively in certain groups. Most of the major chronic diseases occur in all racial or ethnic groups without remarkable differences in frequency. Some of the manifes-

tations of these diseases may differ, however—for example, death from
high blood pressure is twice as likely to be from strokes as from heart attacks among Japanese, Chinese, and Siamese, whereas the reverse holds
in the United States. Tables 12.5 and 12.6 show racial differences in the

Table 12.5. RACIAL AND ETHNIC DISEASE:
SIMPLY INHERITED DISORDERS [a]

Ethnic Group	Relatively High Frequency	Relatively Low Frequency
Ashkenazic Jews	Abetalipoproteinemia Bloom's disease Dystonia musculorum deformans Factor XI (*PTA*) deficiency Familial dysautonomia Gaucher's disease Niemann-Pick disease Pentosuria Spongy degeneration of brain Stub thumbs Tay-Sachs disease	Phenylketonuria
Mediterranean peoples (Greeks, Italians, Sephardic Jews)	Familial Mediterranean fever *G-6-PD* deficiency, Mediterranean type Thalassemia (mainly β)	Cystic fibrosis
Africans	*G-6-PD* deficiency, African type Hemoglobinopathies, especially *HbS, HbC, α,* and β thal, persistent *HbF*	Cystic fibrosis Hemophilia Phenylketonuria Wilson's disease
Japanese and Koreans	Acatalasia Dyschromatosis universalis hereditaria Oguchi's disease	
Chinese	α thalassemia *G-6-PD* deficiency, Chinese type	
Armenians	Familial Mediterranean fever	

[a] SOURCE: McKusick, V. A. 1966. *Mendelian inheritance in man. Catalogue of au-
tosomal dominant, autosomal recessive, and X-linked phenotypes.* Baltimore: Johns Hopkins
University Press. McKusick, V. A. 1967. The ethnic distribution of disease in the United
States. *Journal of Chronic Diseases,* 20:89–101.

Table 12.6. RACIAL AND ETHNIC
DISEASE: MULTIFACTORIAL DISORDERS WITH A COMPLEX
OR UNPROVED GENETIC COMPONENT [a]

Ethnic Group	High Frequency	Low Frequency
Ashkenazic Jews	Buerger's disease (disorder of blood vessels) Diabetes mellitus Hypercholesterolemia Kaposi's sarcoma (a kind of tumor) Leukemia Pemphigus vulgaris (type of skin lesion) Polycythemia vera (too many red blood cells) Ulcerative colitis and regional enteritis (types of intestinal diseases)	Alcoholism Cervical cancer Pyloric stenosis Tuberculosis
Sephardic Jews	Cystic disease of the lung	
Northern Europeans	Pernicious anemia	
Irish	Major *CNS* malformations (anencephaly, encephalocele)	
Chinese	Nasopharyngeal cancer Trophoblastic disease	Chronic lymphatic leukemia Prostatic cancer
Japanese	Cerebrovascular accidents (strokes) Cleft lip palate Gastric cancer Trophoblastic disease	Acne Breast cancer Chronic lymphatic leukemia Congenital hip disease Otosclerosis Prostatic cancer
Filipinos (U.S. only)	Hyperuricemia (too much uric acid—gout)	
Polynesians	Clubfoot Coronary heart disease Diabetes mellitus	
Africans	Ainhum Cervical cancer Esophageal cancer Hypertension	Arteriosclerosis Congenital hip disease Gallstones Major CNS malforma-

Ethnic Group	High Frequency	Low Frequency
	Polydactyly	tions (anencephaly,
	Prehelical fissure	encephalocele)
	Sarcoidosis	Multiple sclerosis
	Systemic lupus orythematosus	Osteoporosis and frac-
	Tuberculosis	ture of hip and spine
	Uterine fibroids	Otosclerosis
		Polycythemia vera
		Psoriasis
		Pyloric stenosis
		Skin cancer
American Indian	Congenital dislocation of the hip	Duodenal ulcer
	Gallbladder disease	
	Rheumatoid arthritis	
	Tuberculosis	
Icelanders	Glaucoma	
Eskimos	Otitis, deafness	
	Salivary gland tumors	

a SOURCE: Damon, A. 1962. Some host factors in disease: sex, race, ethnic group, and body form. *National Medical Association. Journal.* 54:424–431. McKusick, V. A. 1967. The ethnic distribution of disease in the United States. *Journal of Chronic Diseases.* 20:89–101.

incidence of a few selected diseases, for which the observed differences are probably real, and not due to faulty reporting. It is only a partial list. The relative frequencies of these diseases are not specified, although in general the diseases at the top of each group are more closely associated with that population than the diseases at the bottom. Thus sicklemia (hemoglobin S) is virtually confined to blacks in the United States (although it occurs in the eastern Mediterranean), the lipoidoses and pentosuria to Jews, and thalassemia to Mediterranean peoples. Polycythemia vera, a pathological increase in red blood cells, and leukemia, a pathological increase in white blood cells, are about twice as common among Jews as among white non-Jews. Chronic lymphatic leukemia is half as common among Orientals in the United States as among white persons or blacks, who have equal rates. Primaquine sensitivity, or glucose-6-phosphate dehydrogenase deficiency, is manifest in the destruction of red blood cells (as a hemolytic anemia) induced by several drugs, including primaquine (an antimalarial), naphthalene (mothballs), and fava beans, eaten in some Mediterranean countries. This condition occurs in some 13 percent of American black males. Its frequency is higher in some African Negro tribes. It varies between 2 percent and 36 percent of Sephardic (Mediterranean and Asiatic) Jews, and up to 48 percent of some Sardinian groups. It is transmitted on the X chromosome and hence is sex-linked.

Space prohibits considering all the conditions listed. The table has some significant omissions. Being based on prevalence or mortality data, it ignores population differences in the severity or manifestations of disease. For example, not only is the prevalence of pulmonary diseases greater among black and Mongoloid Americans than among white Americans, but the course tends to be more severe among the nonwhites. The hemolytic anemia of primaquine sensititivy is more severe and is induced by more different drugs among sensitive Caucasians than among sensitive Negroes.

Accounting for the data in Tables 12.5 and 12.6 is beyond the scope of this chapter. Some of the differences are known to be determined by single genes; others are probably genetic, but with multiple determinants; all of the genetic manifestations are influenced by the total environment of the individual, both genetic—in the form of modifying genes—and exogenous. Ethnic differences in customs and living conditions probably play the chief role in cervical cancer, tuberculosis, and alcoholism, and a major role in breast and stomach cancer.

I published some of these data in 1962 to draw scientific attention to the racial and ethnic distribution of disease. Only five years later, enough knowledge had accumulated to permit McKusick to expand greatly this work, and to present two much larger tables. Table 12.5 shows single-gene conditions and Table 12.6 shows the racial or ethnic distribution of diseases of complex genetic origin or of incompletely understood origin, only in part genetic. In both tables the large number of diseases, and the diversity of racial and ethnic groups involved, are notable. This massive evidence should persuade us that race is a valid object of biological study, and that racial differences do not stop at the epidermis.

SUGGESTED READINGS

Coon, Carleton S. 1962. *The origin of races.* New York: Knopf.
Damon, Albert. 1969. Race, ethnic groups and disease. *Social biology.* 16:69–80.
Garn, S. M. 1971. *Human races,* 3d ed. Springfield, Ill.: Thomas.
King, J. C. 1971. *The biology of race.* New York: Harcourt.
Osborne, R. H., ed. 1971. *The biological and social meaning of race.* San Francisco: Freeman.
Spuhler, J. N. and Lindzey, G. 1967. Racial differences in behavior. *Behavior genetic analysis.* Hirsch, J., ed. New York: McGraw-Hill.

13
CONSTITUTION AND BODY FORM

The fourth dimension, or kind of human variation, is body form. It is one that is very obvious and familiar to us, but it is the least important, ranking after age, sex, and race. Biological anthropologists generally study man in the mass; that is, groups of normal people. In this they differ from clinical scientists, such as physicians who focus on pathological individuals who are diseased. In recent years, however, anthropologists have become interested in the individual as well as in the group—and one of the meanings of *constitution* is as a shorthand expression for *anthropology of the individual.* Most research in constitution has come from medical problems, and most of its applications have been directed toward disease. Accordingly, much of this chapter is necessarily oriented toward disease, although there are many other applications of constitution, as we shall see.

Definitions

Constitution, which literally means "standing together," is the sum of person's innate, relatively fixed biological endowment. In relation to disease, constitution represents the host factor in the host-agent-environment complex. George Draper, in the 1920s, regarded the individual as a four-sided pyramid resting on a base (Figure 13.1). The four sides are the anatomical, biochemical, immunological (i.e., disease-susceptibility), and the psychological aspects of the person, while the base is his genetic makeup. It is these aspects of the individual that "stand together," and their in-

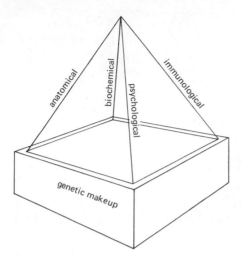

13.1 Human constitution may be envisioned as a four-sided pyramid on a base of man's genetic makeup.

terrelationships comprise the subject matter for the discipline of constitution.

The anatomical aspect, chiefly body build, is but one of several constitutional approaches, and less important than the other constitutional factors of age, sex, race, biochemical and genetic makeup, and disease susceptibility. The morphological aspect of constitution has, however, been the traditional approach, partly because it works—that is, physique actually does correlate with disease and behavior to some degree for historic reasons; and partly because physique is obvious and easily described. *Constitutution* means different things to different disciplines. To a clinician, constitution means his patient's biological individuality. To an epidemiologist, it means host factors in disease. To an immunologist, it means tissue specificity, and to a blood specialist constitution means transfusion reactions. To physical anthropologists, psychologists, and behavioral scientists generally, constitutiton means physique in relation to environmental adaptation, disease, or behavior. As such, constitution is one application to man of the structure-function relationship. This relationship between structure and functions is a central concern of physical anthropology and for that matter of many other disciplines in the natural as well as the biological sciences.

Reasons for Study

We study body form in relation to disease for the same reason that we study age, sex, race, and other personal or host characteristics. These reasons are:

1. To predict in advance who is susceptible to a given disease and who will respond in what way to disease or to treatment therapy. Response to treatment can be as important as who gets the disease in the first place. Prediction can help to prevent, diagnose, and treat disease. For example, we can now identify those individuals at high risk for coronary heart disease as those combining family history of the disease, tobacco smoking, sedentary habits, high blood pressure and serum cholesterol, and obesity and mesomorphy (fullness of build based largely on muscle and bone development).

2. To obtain clues to the mechanisms underlying any associations that may be found. Such clues may be sought in several areas: in biochemical and physiological correlates of disease; in anatomical relationships such as the distribution and structure of organs, tissues, and cells correlated with external body form; and finally, in genetics, via pleiotropism.

3. The associations found between body form and function will suggest certain laboratory investigations, which may help to elucidate the causes of disease. It is obvious that body build does not itself cause disease. What then is the common factor that links them?

4. The same process can disclose several places in the web of causation at which intervention can help prevent disease. For example, now that many stages of cholesterol metabolism have been demonstrated, one can intervene at many points in order to prevent coronary heart disease.

It is of course possible to intervene without understanding all the causal mechanisms. Identification of individuals and groups at high risk, by the use of constitutional as well as other criteria, will permit environmental manipulation. For example, if stocky men, both muscular and fat, are particularly prone to early coronary heart disease, which seems to be the case, such men can be examined frequently and advised as to their diet, exercise, and smoking habits—aspects of their environment which also increase the risk, but which are subject to voluntary control.

Even if one cannot prevent a disease, one may hope to postpone it. A good case can be made that one should not try to eradicate coronary heart disease. Since everyone must die of some disease, one would search a long time before finding a better cause of death. It is sudden and virtually painless, in great contrast to protracted, painful, and expensive diseases such as cancer or stroke. What we really hope to achieve is to postpone the disease as long as possible. Currently, many men get heart attacks in their forties and fifties at the height of their occupational powers and when they are most needed by their families. Nobody would object to dying from a heart attack like the retired Lord Chief Justice of Britain, who expired from excitement just after landing a spirited salmon at the age of 95.

Some Historical Notes

On the varied background of constitutional medicine one finds folklore, liteterature, and science. The first two mix easily. Whether or not the mean Scrooge was thin and scrawny, that is how he is drawn, in contrast to hearty, jolly, fat Falstaff. And what Shakespeare put into the mouth of Julius Caesar has been quoted over and over again:

> Let me have men about me that are fat,
> Sleek-headed men, and such as sleep o'nights.
> Yond Cassius has a lean and hungry look.
> He thinks too much, such men are dangerous.
>
> .   .   .
>
> Would he were fatter! . . .
> I do not know the man I should avoid
> So soon as that spare Cassius.

Scientifically and medically these ideas go back to Hippocrates, who distinguished the same body forms: phthisic, or tuberculosis-prone, and apoplectic, or stroke-prone. This comes down through the European tradition of astute observational diagnosis (as against bubbling tubes and computers). The great English surgeon, Sir William Osler, said that it is as important to know what kind of person a disease has as the other way around. (Osler may be due for a revival among junior academics for his more famous suggestion, that all men should be chloroformed at sixty—which he actually borrowed from Anthony Trollope.) The American, George Draper, distinguished gallbladder and ulcer "races."

In addition to medicine and epidemiology, the constitutional approach has made its way into psychiatry and criminology. In the first, the German Ernst Kretschmer set up pyknic (heavy), athletic (muscular), and asthenic (lean) types, associated with different psychoses, later followed in this country by William H. Sheldon, who converted these types into axes of variation (see below). In criminology, the famous Italian Cesare Lombroso believed he had established criminal stigmata, or physical marks of the criminal type, followed by E. A. Hooton, who discerned slight statistical differences among criminals by type of crime, and by

Sheldon and Eleanor Glueck, who found juvenile delinquents to vary in a mesomorphic direction.

In medicine, after a period of down-grading at the hands of environmentalists and behaviorists, the host factor has come back toward the center of the stage, as the degenerative diseases of age have more and more replaced the infectious diseases of youth with the stamping out of so many of the latter.

Constitutional studies have had more than their share of fakes and charlatans, in the persons of the mantics and phrenologists, as well as their respected scientists. In fact Sherwood Washburn, an eminent anthropologist who is unsympathetic to constitutional studies, has warned that nowadays we are merely substituting "bumps on the buttocks" for bumps on the head. I hope to show that this is not the real state of affairs.

A fascinating study in intellectual history, not to say irrational behavior, would be the history of constitutional medicine and its current lowly standing in the social and behavioral sciences, in contrast to its respectable standing in the medical sciences. Three generations of social scientists have now made careers by debunking Lombroso, Hooton, and Sheldon. But the recent discoveries that some violent criminals have the XYY chromosome constitution, while others have specific brain disturbances have restored scientific respectability to the old folk notion that there may be some biological component in violent behavior. It is interesting that the obvious correspondence between structure and function is readily acknowledged and freely studied in the chemical molecule, the cell, and in organisms below man, but when we come to man, what should be an area for rational investigation may in some quarters arouse violent emotion and heated denial.

Methods of
Analyzing Physique

Classical Anthropometry. This field consists of body measurements (Figure 13.2). These are still useful—in fact, height, weight, and their relationship are still the most important general measurements of physique, not only for describing body form but also for relating body form to function (Figure 13.3). Such functional aspects include growth, as in the growth standards used by pediatricians; nutritional status; and effects of disease. Height, weight, and skinfolds are far more sensitive indicators of malnutrition in an individual or a group than are medical or biochemical

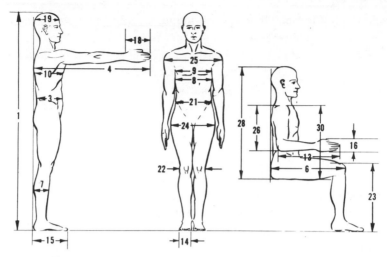

13.2 Some human body measurements with a practical application in the field of vehicle design. Measurements such as these enable designers to plan the interior spaces of cars, trucks, buses, and airplanes so that drivers and pilots may drive more efficiently with less body fatigue.

13.3 The controls of this truck were designed with little thought of the needs of the people who would operate it. More than comfort is involved in human engineering; highway and air safety are also involved.

examinations. One relates these measurements of an individual or a group to standards based on well-nourished persons of the same genetic background. Such simple body measurements can indicate the need for dietary supplementation, whether for an individual child or for a segment of a population. Once such dietary programs have been introduced, one can measure their effect, or lack of it, by the same simple body measurements.

Body dimensions are also useful for other functional purposes. For example, height and chest dimensions correlate with lung function and lung diseases. Limb length as a percentage of total height is altered in glandular and chromosomal disorders. Limb length is increased in eunuchs, hyperpituitarism (gigantism), and in the XXY (Klinefelter) syndrome; it is decreased in cretins, achondroplastic dwarfs, and the XO (Turner) syndrome. Measurement of limb circumferences is useful as a guide to blood clots or to neuromuscular, bone, or joint disease.

The drawbacks of classical anthropometry are first, its traditional content, devised one hundred years ago and oriented toward the skeleton and toward evolutionary and racial problems; secondly the training required to achieve technical and professional competence; and finally, the multiplicity of possible measurements and the difficulty in interpreting results. Using measurements alone, Ales Hrdlička of the U.S. National Museum was unable, in the 1920s, to give the U.S. Navy a simple classification of body build, for selection of divers. As we shall see below, Sheldon could—endomorphs should not become divers or flyers because they are more subject to aeroembolism.

Indices. These are the ratio of one measurement to another, and extreme values of ratios of one measurement to another are viewed as *disproportions.* The technique of application to constitution was devised by Carl Seltzer: disproportions reduce the many possible measurements to a few presumably key ratios. Seltzer's reports, based on Harvard Grant Study subjects, showing that superior performers within an occupation had fewer disproportions than average performers, were confirmed among military flyers but not among civilian bus and truck drivers.

Factor analysis. This is the attempt to find a small number of factors, whether independent or correlated, which account for the correlations between many body measurements, taken in pairs. Just as psychologists look for primary mental abilities from many test results, anthropologists look for primary factors or clusters of physique from many body measurements. Factors of human physique identified so far include one for general size and one for linearity versus laterality of build, but beyond that there is little agreement. Moreover, the factors of physique that have been identified have rarely been correlated with any aspect of behavior—in fact, the only worker to attempt this to my knowledge found that physical factors correlated no more closely with athletic performance than did somatotype ratings (see below).

In any case, both indices or disproportions and factor analysis are highly sophisticated research tools still under development, with limited or even questionable accomplishments to date. They by no means fill the need for a simple, clinically feasible method for describing physiques.

The next two techniques attempt to distinguish between the fat and the bony-muscular components of body build. All three preceding methods easily identified a linear-lateral, or lean-stocky axis of physique, which has proved useful in correlating with disease and behavior, as we shall see. The following techniques, body component analysis and general ratings, or types of body build, try to refine stockiness of build still further into fat, bone, and muscle.

Analysis of Body Composition, or Body Tissues. Such analysis attempts to partition the body into its percentages of water, fat, and lean body mass—namely bone, muscle, and connective tissue. Among the techniques used are skinfold measurement, densitometry, volumetrics, radiography, and dye or isotope dilution and distribution. While all of these methods involve appreciable error, it is fortunate that the use of two simple skinfolds permits prediction of total body fat among young men to within 2 percent, the range of error of densitometry. Body composition analysis has been applied to medicine for several purposes: to refine estimates of obesity, to derive somatotype ratings in the absence of photographs, and to estimate the severity of disease after the disease has occurred. Except for skinfold measurement, which is an anthropometric technique, the analysis of body composition remains a laboratory exercise rather than a clinical or field science.

Rating of Body Form. Beginning with the Hippocratic phthisic and apoplectic habitus, rating of body form has had a long history. The two Hippocratic types gave way to three, as in Kretschmer's pyknic, athletic, and asthenic types. Sheldon has retained these three dimensions, which he calls endo-, meso-, and ectomorphy (Figure 13.4), and has added a fourth, gynandromorphy, or morphological masculinity versus femininity.

Sheldon made two major contributions: he replaced polar types with continuously varying dimensions of physique, and he introduced objectivity in the form of standardized photographs and an age-height-weight basis for ratings. Sheldon's somatotype system, despite many unanswered questions concerning the stability of somatotype with age and under varying conditions of nutrition, and the biological interpretation of the components, has provided a useful description of physique, as we shall see.

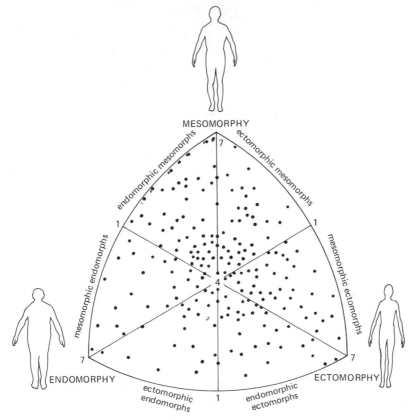

13.4 Human somatotype: mesomorphy, endomorphy, and ectomorphy, a schematic two-dimensional projection of the theoretical spatial relationships among known somatotypes. Illustrated is the distribution of a male college population of four thousand. Each dot represents twenty individuals. Note the randomness of the distribution. The figures that have been sketched in are rather extreme expressions of each type.

All the current efforts to establish relationships among the various anthropometric techniques seek a simple method which is not photographic and is clinically feasible. This means nonimmersion, nonradiographic, and noninjection. There should be an objective and understandable battery of a few external observations which will (1) describe body form, (2) correlate with body composition, and (3) relate to behavior, physiology, and disease (Figure 13.5). Since all the systems now in use, including standard anthropometry, have shown associations with some human function, the multiplicity of methods for describing physique should not obscure the fundamental fact that such associations do exist. In man, as elsewhere in nature, form and function are closely related.

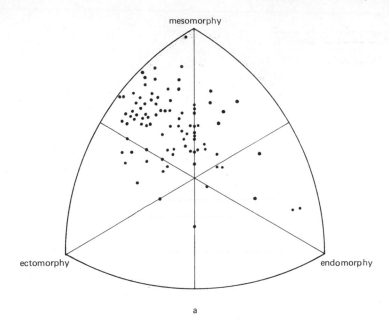

a

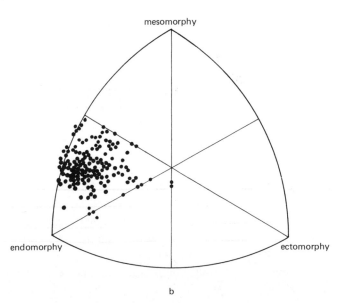

b

13.5 The distribution of somatotype among ninety-two men having duodinal ulcers (a) and one hundred forty-six diabetic women (b). (Data, from author's studies made at Presbyterian Hospital, New York City.)

Results and Applications

Let us consider applications to normal anatomical and physiological standards. Age-height-weight standards for insurance, military, and industrial purposes are increasingly taking note of differences in body build as modifying the so-called normal or desirable weights. Insurance companies now qualify their recommendations in terms of large, medium, or small frames. The military services have learned through their rejection in World War II of many outstanding athletes as overweight, that weight can reflect muscle as well as fat. We may hope that skinfold measurement will become a standard part of the examining physician's routine (Figure 13.6).

Anatomically, correlations have been established by Bjurulf in Sweden between gross external anatomy, or body form, and cellular morphology. He reported that the number of fat cells was mainly genetic in origin, whereas the size of fat cells was mainly environmental, varying with age and state of nutrition. Bjurulf is currently extending this work to muscle. Very recent experimental work suggests that early overfeeding in animals increases the number of fat cells, which are never lost thereafter. So it is possible that our custom of stuffing infants with food promotes the obesity that afflicts so many of us in later life.

Physiologically, external body form has been correlated with specific

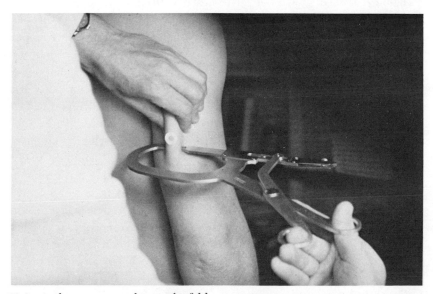

13.6   Anthropometry: taking a skinfold measurement.

gravity, which is a measure of fat. Specific gravity correlates highly with skinfolds (0.70) and with endomorphy (0.85). Blood pressure varies with laterality of build (that is, stockiness or mesomorphy) rather than with obesity. Lung function is independent of somatotype, but lung capacity is closely correlated with stature—oddly enough, more closely correlated with stature than with sitting height, a measure of the trunk without the legs. As for strength, forearm strength correlates well (0.57) with forearm circumference, so that forearm girth becomes a useful index of strength and muscularity. With respect to growth, mesomorphs reach maturity— whether dental, skeletal, and sexual—earlier than do ectomorphs. But, as adults, early maturers close their epiphyses earlier, when their sex hormones have reached adult levels of production. On the other hand, slower maturers, whether males as compared to females, or ectomorphs as compared to endomorphs, have a longer growth period, so that they may eventually equal or exceed the height reached by the early maturers.

Genetically, physique studies of twins have shown that body lengths are chiefly genetic in origin; breadths and girths environmental. Body build or somatotype, particularly ectomorphy, is largely genetically determined as are certain measures of muscle (upper arm girth) and bone (wrist breadth).

Biochemically, stockily built men have higher serum cholesterol levels, as expected from the increased risk of coronary heart disease run by stocky men. They also have higher serum uric acid levels and gout. An association has been shown between linearity of build and another disease precursor, serum pepsinogen level. Both linear men and those with elevated serum pepsinogen have an increased risk of developing peptic ulcer. So all three of these associations point in the same direction: the same kind of people—not necessarily the same individuals, but those of generally similar body build—who show the biochemical abnormality also tend to develop the disease associated with that biochemical abnormality.

As for other biochemical associations with body build, James Tanner has reported high correlations between creatinine, which is an index of muscle mass, and weight (0.76), surface area (0.63), skinfolds (0.47), and somatotype (0.44–0.49). Among men, androgen (male sex hormone) levels vary with mesomorphy and grip strength, but in a group of 244 men whom we studied we were disappointed to find no relationship between these male sex hormone levels and gynandromorphy, or masculinity of physique. Normal women have not yet been studied in this respect, although it is interesting that women with cancer of the breast or cervix, which are hormone-dependent cancers, tend to be hyperfeminine in body build.

Physique and Behavior. In the field of behavior, aside from folklore and literature, scientific interest in the body-behavior relationship has been of long standing. It can now be said that real associations do exist, for both normal and abnormal behavior. Sheldon in 1942 reported high correlations between somatotype and temperament—that is, the twenty traits that we have seen make up each of his three components of temperament. Nobody else has confirmed this—partly because nobody has tried, but possibly also because Sheldon made both sets of ratings himself, physique and temperament. Actually, associations have been reported between physique and occupational selection and success among military pilots, soldiers, industrial workers, bus and truck drivers, lumberjacks, and students in different fields of concentration. The relationship of occupation to disease, particularly coronary heart disease, is pertinent

Table 13.1. PHYSIQUE AND DISEASE [a]

DISEASE	SUSCEPTIBLE PHYSIQUE
I. Disputed Association, Probably Real	
Peptic ulcer	Linear
Gallbladder disease	Stocky, lateral, fat
Psychoses: schizophrenia	Slender, lightly muscled, (asthenic)
manic depression	Stocky, short, powerful (pyknic)
II. Reported but Unconfirmed Association	
Poliomyelitis	Mesomorphic
Meniere's disease	Mesomorphic
Otosclerosis	Ectomorphic
Breast and cervical cancer	Hyperfeminine
Uterine fibroids	Mesomorphic
Hypertension, malignant	Lean
III. Confirmed Association	
Tuberculosis	Tall, thin
Diabetes: adult	Obese
juvenile	Tall
Cancer, endometrium	Obese
Hypertension, essential	Stocky, lateral
Coronary heart disease	Endomesomorphic

[a] From various sources.

here, in that the same physical types prone to coronary heart disease, the stocky endo-mesomorphs, predominate among a high-risk occupational group, bus and truck drivers. Delinquent boys have been more mesomorphic than nondelinquent controls.

With regard to habits of smoking and drinking, tobacco smokers have been reported to differ physically from nonsmokers, independently of food intake, and even before beginning to smoke. Unfortunately, the direction of difference was opposite among young men and adults: young smokers were heavier than nonsmokers, whereas adult smokers were lighter and leaner than nonsmokers. This discrepancy is still unexplained.

Alcohol consumption seems to be independent of physique, but it is possible that constitutional research may eventually disclose persons who should avoid certain drugs, foods, or occupations.

A final application of constitutional studies to one kind of behavior is in the area of reproduction. So far, among completed families of Harvard men, stocky men produced a higher proportion of sons than lean men, but the total number of children was similar for men of all body builds. Men in the mid-range of height had more children than the very short or the very tall.

Physique and Disease. The relations of body form and disease are best shown in tabular fashion (see Table 13.1). While the associations reported vary in strength, there are indeed real associations for certain diseases; above all: coronary heart disease, diabetes, gallbladder disease, stroke, cancer of the uterus, tuberculosis, and peptic ulcer.

Current Research in Constitution

Improvements in Technical Methods and Objectivity. To eliminate two requirements of the standard procedure—photographs and subjective ratings of body—statistical combinations of body measurements can be formed to give somatotype ratings. In addition, such statistically combined measurements are easier to relate directly to the outcome of interest, whether a biochemical determination or the occurrence of a disease. This completely bypasses the use of a somatotype, a thing which seems to trouble many scientists because of its subjective aspects. The statistical technique I am describing is known as multiple regression. With its use, it has proved possible to predict, with some success, which middle-aged men, healthy at the time of examination, were more likely to develop coronary heart disease within the next decade.

Relations of Physique to Physiological Function and Microscopic Anatomy. A second active area of research is of the kind lately conducted by Bjurulf, relating body build to cell and tissue structure. It is possible, for example, that stocky mesomorphs have their recognized greater risk of developing coronary heart disease because they have a distinctive structure of the arterial walls. This is a very promising lead.

Mechanisms Underlying Association. A final area is a still further pursuit of associations of body build and biochemistry. In this area one asks what mechanisms are responsible for these different manifestations of human constitutional individuality? The answers to such questions promise large dividends in detecting, treating, and preventing disease, by pointing to the target at which intervention should be shooting.

SUGGESTED READINGS

Bauer, J. 1963. *Constitution and disease,* 3rd ed. New York: Grune and Stratton.

Damon, Albert. 1961. Constitution and smoking. *Science.* 134:339–341.

Damon, Albert. 1900. Some host factors in disease: sex, race, ethnic group and body form. *American Journal of Physical Anthropology.* 22:375–382.

Damon, Albert, Bleibtreu, H. K., Elliot, O., and Giles, E. 1962. Predicting somotype from body measurements. *American Journal of Physical Anthropology.* 20:461–474.

Dunbar, Flanders. 1954. *Emotions and bodily changes.* New York: Columbia University Press.

Kretschmer, E. 1936. *Physique and character.* London: Kegan Paul.

Osborne, P. H., and de George, F. V. 1959. *Genetic basis of morphological variation.* Cambridge, Mass.: Harvard University Press.

Sheldon, W. H. 1940. *The varieties of human physique.* New York: Harper.

Sheldon, W. H. 1953. *Atlas of men,* New York: Harper.

Sheldon, W. H. Constitutional variation and mental health, in *Encyclopedia of mental health.* New York: Franklin Watts.

14
NATURE AGAINST MAN: LIGHT, HEAT, COLD, HEIGHTS

In considering man's response to his environment, anthropologists and physiologists differ in several respects. A physiologist is interested in the short-term response of individuals to a single kind of stimulus under carefully controlled laboratory conditions. An anthropologist, on the other hand, studies long-term adaptations of whole populations to their own natural environments. Human variation is an embarrassment or a nuisance to the physiologist, who, like the engineer, the chemist, or the physicist working with uniform and controlled materials, tries to define and measure human response in a way that will hold for the whole species. Often he succeeds. In fact, Nobel Prizes in physiology have been won, and deservedly so, for a critical experiment on one or two subjects, such as the effect of the nervous system on the heart, or the effect of certain hormones or drugs on the heart or kidney. But human variation itself is what the anthropologist studies and tries to explain in terms of long-term adaptation of populations to their differing natural environments. In this chapter and the next I shall try to deal with the major environmental influences that have shaped man so far and the newer ones that man is creating for himself and that will guide his present and near-term adaptation.

Race and Adaptation

We have already discussed the major geographic populations, or races of man, in terms of physique and biochemistry, and of differential susceptibility to disease. Let us now examine some notions of how races got that way. As we have seen, it is clearly impossible to account for all the morphological and biochemical differences among races by simple Darwinian adaptation or selection. [Figures 14.1–14.12.] Nobody knows, or can even hazard a guess, as to the adaptive significance of lip thickness, tooth size and shape, hair form, ear size and shape, or even some of the blood group or biochemical differences. There are several possibilities: (1) distinctive racial traits may be or may have been useful at one time, in ways still to be determined; (2) distinctive racial traits may be completely neutral and may have arisen and become maintained by pure chance— i.e., mutation and drift; (3) distinctive racial traits may represent pleiotropic effects of other genes which are more directly useful. Boyd thinks the last is the answer. I have mentioned earlier two further possible mechanisms—sexual selection, for loss of body hair and other more specifically racial features; and another mechanism, end-organ failure to respond.

In the present state of our ignorance, rather than proceeding trait by trait or even race by race, it seems better to take broad ecological areas of the world, such as the tropics, desert, and arctic environments, or broad physical characteristics, such as body build or pigmentation, or broad types of physiological stress and adaptation, such as heat, cold, altitude, and nutrition, and see if we can observe any regularities corresponding to the racial differences which exist today.

Skin Color

The environmental variables most relevant to population differences are solar radiation, temperature, nutrition, and disease. Deeper pigmentation, or skin color, protects against damaging ultraviolet radiation from the sun, which causes sunburn and skin cancer. The visible spectrum of light ranges from violet, with a wavelength of 3800 Å (ångstrom units), to red, at 8000 Å. Light beyond the violet, or ultraviolet light, is invisible, but in the range of 3000 to 3200 Å it will cause sunburn. Man has three kinds of skin pigments: carotene, a yellow pigment which we obtain by eating foods such as carrots and sweet potatoes; hemoglobin, which is the red oxygen-carrying pigment of red blood cells; and melanin, a complex brown pigment produced in the basal layers of the epidermis, already described. Skin color depends on the relative proportions of these three pigments. Of the three, only melanin absorbs ultraviolet radiation and thereby protects against sunburn and skin cancer. In the tropics, albinos, who lack melanin, develop skin cancers in early childhood. On the other hand, heavily pigmented people have virtually no skin cancer. In the United States, Caucasians in Texas have five times as much skin cancer as in Massachusetts. Incidentally, people of all shades tan, except for albinos.

Vitamin D. Another effect of sunlight on man, particularly ultraviolet radiation, is to activate certain fats (steroids) in the skin, notably dehydrocholesterol (ergosterol), to form vitamin D. Vitamin D is necessary for proper bone growth; people who lack it develop rickets, with bowed legs and other defects. People who have seen little sunlight, such as slum populations in Britain during the worst of the Industrial Revolution, or upperclass women in traditional Chinese society, who never left their homes, might develop rickets. In the higher northern latitudes, winter sunlight is weak, and in Europe the skies are indeed cloudy all day. Here it is obviously desirable to absorb as much sunlight as possible. The best way to do this is to have a light skin, with reduced amounts of melanin and very little screening out of ultraviolet light.

Color and Latitude. Man's original skin color is unknown, but we do find a general parallelism between skin color and solar radiation, from one major race to another; and even within each major race people living near the equator are darker than those near the poles. Over the whole Caucasian range, in Europe, North Africa, and India, skin color is darkest nearer the equator and lightest farther away. As already noted there are some exceptions to these generalizations, notably the absence of pigmentary differences within the American Indians despite their having occupied an extreme range of habitats running nearly from pole to pole for twenty to thirty thousand years. Also, early European settlers found deeply pigmented Tasmanian aborginals living in a cool, cloudy climate, much like that of England or New England. For lack of a better explanation, we say weakly that 20,000 years may be too short a time to develop such differences, but 600 generations would seem to be enough to have produced something.

Gloger's Rule. I must also mention one of the three famous Rules of Animal Distribution, or Zoogeography. Gloger's pigmentation rule states that mammals and birds living in warm, moist climates have darker skin

14.1–14.4 Caucasoids.

14.1 (Left) Northern European woman. (courtesy Warder Collection)
14.2 (Right) Ainu, northern Japan. (courtesy Peabody Museum, Harvard University)

14.3 (Left) Kandian chiefs, Sri Lanka. (courtesy Peabody Museum, Harvard University)
14.4 (Right) Arab family. (courtesy Warder Collection)

14.5–14.8 Africans.

14.5 (Left) Bantu woman. (courtesy Peabody Museum, Harvard University)
14.6 (Right) Bushman, southwest Africa. (courtesy Peabody Museum, Harvard University)

14.7 (Left) Pygmy of the Congo Forest, Africa. (courtesy Peabody Museum, Harvard University)
14.8 (Right) Masai woman, Africa. (courtesy Peabody Museum, Harvard University)

14.9–14.12 Southeast Asians.

14.9 (Left) Bhutan woman. (courtesy Peabody Museum, Harvard University)
14.10 (Right) Balinese man. (courtesy, David Irons)

14.11 (Left) Borneo man. (courtesy Peabody Museum, Harvard University)
14.12 (Right) Vietnamese men. (courtesy Peabody Museum, Harvard University)

or coloration, where as those in arid regions are yellow or red-brown. One can find agreements in man, such as the South African Bushmen, but also disagreement, such as the Australian aborigines.

While heavily pigmented skin protects against ultraviolet radiation, it also absorbs more heat, and hence produces an extra load in hot climates—and hot climates in general are those with high radiation. (There are exceptions, as always: Arctic areas are cold, but they involve much solar radiation, both direct and reflected, while African equatorial forests are shady.) A dark skin was once thought to be advantageous in hot climates, since the extra heat absorption stimulates early sweating. But because the extra heat load persists throughout the exposure this is too high a price to pay for a dubious initial advantage. The untanned black skin absorbs 34 percent more solar energy than does white skin. This additional 34 percent can indeed impose a great strain on the body's heat-regulating mechanisms, especially sweat production. In fact, Paul Baker found that on the Arizona desert, among a sample of forty pairs of black and white soldiers matched for age, size, and body fat, the blacks developed higher skin and rectal temperatures than the whites.

Although blacks are worse adapted to desert or dry heat, they tolerate moist heat better in the sense that, when fully acclimatized, they perform work with less sweat production than whites. During sweating, the body loses salt and water, which should preferably be retained. The demonstrably better adaptation of blacks to moist heat than to dry heat has led to the speculation that dark skin is adaptive not so much to sunshine as to humid tropical forests, where it affords concealment. This remains a pure speculation so far.

Body Size and Shape

On the whole, body size increases as one proceeds from the equator, both north and south, to cooler regions. The best evidence for this is Derek Roberts's work. His correlations were based on published anthropometric series that happened to be available. These series were heavily weighted, as it happened, by groups which showed the negative association between body size and temperature. For example, Roberts included many African pygmy groups (living along the equator) but only two Polynesian groups— and Polynesians, who also live near the equator, are notoriously tall and heavy. Lapps, near the North Pole, are short and slender. A few more

Polynesian groups, some Lapps, and more Melanesians might have given a very different picture.

Bergmann's Rule. Despite this warning, the association seems to hold, not only among different peoples but also within single continental areas. Since weight, rather than stature, increases as one proceeds from the equator north or south, body mass increases relative to surface area. This illustrates Bergmann's body mass or body size rule: "In otherwise similar bodies, the larger one has the smaller skin surface in proportion to mass, since volume and mass increase as the cube of the linear dimensions, and surface only as the square."

Allen's Rule. Obviously, the above arrangement is an advantage in retaining metabolically produced heat, since heat is produced in proportion to mass and is lost at the surface (by radiation, conduction, and convection) in proportion to surface area. Laterally built bodies, with short extremities, as in Eskimos and many other Mongoloids, also exemplify Allen's rule, which notes that in colder climates animals have smaller appendages and extremities. The elephant's ears plus trunk are said to illustrate Allen's principle, providing surface area for the radiation of heat. But what about the rotund rhinocerous?

How are these observed differences in body build, associated with climate, to be interpreted? In general, the tropics are also zones of poor nutrition, parasites, and other disease, so that alternative explanations for small body size, besides the climatic one, are possible. Here the statistical technique of multiple correlation can help decide. Both Roberts, for worldwide distributions, and Marshall Newman, for the Americas, found that stature and temperature together account for some 80 percent of the variance in weight, leaving only 20 percent to be accounted for by other factors, such as genetics, nutrition, and disease. Climate therefore does indeed seem to be the major regulatory factor for human body size and proportion.

Subcutaneous Fat

Since fat insulates, one would expect to find more fat among people who live in colder environments. Actual studies, however, show a rather different picture. In A. Steegmann's experiments facial fat, as in Mongoloids, did not protect against cold, and he points out that facial frostbite is rare and mild in any race! Among Western peoples, food and exercise are the important factors in adiposity. Middle-aged urban white Americans have the highest proportion of subcutaneous and total body fat recorded for any population: at age 55, 26 percent for men, 38 percent for women. At the same ages, Spanish laborers had about half the skinfold thickness of Spanish professional men. Among ten diverse groups reported by Elsner, South African Bushmen had the least skinfolds, 4.1 mm (average of eight sites); Alaskan Eskimos, 5.7; Canadian Eskimos, 5.8; Arctic Indians, 6.2; Quechua Indians (Peru), 6.5; Australian aborigines, 7.0; Lapps, 7.2; Alacalufs, 7.9; citified Australian aborigines, in Darwin, 9.1; urban white Australians, 9.6 mm. Adult male Solomon Islanders, for comparison with American men aged twenty years and older, showed the following means for skinfold thickness (mean of triceps and subscapular skinfolds): Nasioi, 7.1 mm; Kwaio, 7.1; Lau, 7.7; Baegu, 7.3; U.S., 14.0. Note that these various differences do not follow racial lines.

Cold Adaptation. Short-term experiments on Caucasians, using a standardized sleeping-bag test in the cold, have shown that fat helps insulate the body; the fatter subjects retained higher internal body temperatures. But this did not hold for various populations studied in the same way by Elsner. These groups showed no correlation between skinfold thickness on one hand and rectal temperature, average skin temperature, metabolic rate, or body tissue conductance on the other. The several groups adapted to cold in different ways physiologically, not anatomically. These were the results.

1. Unacclimatized urban Caucasians under cold stress had falling skin and rectal temperatures, rising metabolic rates, shivering, and wakefulness.

2. The uncivilized groups all slept better, and showed three types of response: (a) Metabolic acclimatization: Arctic Indians, Eskimos, and Alacaluf (Tierra del Fuego) had high metabolic rates and warm extremities. (b) Insulative acclimatization: Central Australian aborigines and Kalahari Bushmen had stable or falling metabolic rates and cooler skin. (c) Peruvian Quechua Indians from high altitudes, and Lapps had low rectal temperatures and maintained high temperatures in their extremities, i.e., they redistributed their body heat. This is hypothermic or redistributive acclimatization. (d) Finally, acclimatized Norwegian students resembled the Eskimo and Alacaluf—metabolic acclimatization.

Since the Australian aborigines and South African Bushmen live in climates where there are materials for fire at night, where temperatures below freezing are rare, and where cool winter nights are followed by warm sunny days, cold injury is unlikely. Their distinctive response to

cold, namely peripheral vasoconstriction in the absence of much subcutaneous fat, can therefore be useful in the given conditions. High metabolic rates would be a disadvantage, since these desert dwellers are also
heat adapted. Their food supplies are limited, and might not be able to
sustain a high *BMR* (basal metabolism rate). This insulative adaptation is
the only distinctive form of cold adaptation for man, who is originally a
tropical animal. All other groups adapt to cold by attempting to warm the
extremities by peripheral vasodilation; the Eskimos are the best at this
hunting phenomenon. Africans show the least rise in *BMR* and the coldest
extremities and are therefore least adapted to cold. This conclusion is sup-
ported by the relatively high frequency of frostbite among American black
soldiers during the Korean War.

It is possible that melanin in some way increases the susceptibility of
the skin to cold. In one experiment, two samples were taken from single
piebald (black and white) guinea pigs subjected to cold temperature.
From the same animals, skin samples containing melanin showed much
greater blistering and cell destruction than skin samples without melanin.

It is easy to see that people living in extremely cold temperatures,
like the Eskimo and the Alacaluf of Tierra del Fuego, cannot afford to
lower the temperature of their hands and feet. This is why they adapt to
cold by increasing their heat production and developing efficient physiological mechanisms for warming their extremities. We may therefore conclude that response to cold is physiological rather than anatomic—in other
words, functional rather than structural.

Not only is there no clear correlation of skinfolds with climate, as just
noted, but doubt has been cast on some of the oversimplified notions of
the insulative or other functional value of fat. Wyndham even doubts the
validity of the physiological adaptations. We have already mentioned that
the notion of steatopygia as fat storage for periods of starvation does not
hold; it is not drawn upon in such periods, and probably represents sexual
rather than natural selection.

Racial Differences. Further studies on whites and blacks in the
United States have failed to confirm R.W. Newman's 1956 report that
black Army recruits (mean age 20.8 years) had less subcutaneous fat than
white recruits. My coworkers and I found virtually no difference in skinfold thickness among white and black soldiers aged 24 and 27 years,
respectively, while Comstock and Livesay found middle-aged blacks to
have slightly larger skinfolds than middle-aged whites in Georgia. Other
evidence (e.g., increased skinfold thickness with length of service among
Turkish soldiers) all points in the same direction: namely, that skinfold
thickness—and the body fat which the skinfolds predict fairly

accurately—are chiefly nutritional in origin, when age and sex are equated.

If body weight increases with latitude, that is, as one moves away from the equator to either poles, but skinfolds do not increase, what is the extra tissue? We are left with muscle. This is logical, since muscle is highly active metabolically and can raise the metabolic rate. This rise in metabolic rate, as we have seen, occurs in cold climates and is one way in which man adapts to cold. In fact, physical fitness, which is associated with muscle mass, can increase metabolic acclimatization to cold; and so can diet. Normal Eskimo diet, full of protein and fat, raises the metabolic rate; when they eat a standard Western diet, their metabolic rate decreases.

Heat Adaptation

We have seen that skin color and body build vary with climate. Pigment helps protect against solar radiation, and body build (a linear physique, with relatively long extremities) helps dissipate body heat. This heat loss is accomplished by evaporation of water, some from the lungs via respiration, but most from the body surface. In hot areas, an efficient sweat mechanism is therefore necessary for survival. The evidence on the number of sweat glands in man can be quickly summarized: races do not differ, but individuals do. And within the individual, there is a gradient: upper limb-lower limb-trunk; and within each limb, the portion farthest from the body has more sweat glands than the part nearest to the body. That is, the hand has more sweat glands than the forearm, which has more than the upper arm.

Dry versus Moist Heat. Acclimatization to heat stress is quite rapid, taking only a few days, and occurs in all races. The mechanism is that the sweat glands, acting under the thyroid gland, become more responsive to heat, both in speed of response and in the amount of output. The amount of work a man can accomplish increases, while his pulse rate and cardiac output decrease for a given work load. This rapid acclimatization shows that man was originally a tropical animal. As we have seen, for the tropics blacks are on the whole better adapted to hot, humid environments and whites to hot dry ones, but one must be careful to study large, representative samples of acclimatized subjects, since individuals show considerable variation. Also, long-term acclimatization differs from the short-term situation, in that it is preferable to keep cool and work with less sweating, thereby conserving water and salt. It is on this basis that blacks are said to be better adapted to moist heat.

With regard to dry heat, the most definitive criterion of desert heat tolerance is, of course, survival. Schickele has compiled data on deaths from heat prostration during desert maneuvers in World War II. Height and weight data showed that fatness was a hazard. So also was muscularity, that is, laterality of body build, or the amount of fat-free weight per unit stature. This indicates the value of the linear body build, as in desert Arabs, Nilotic Africans, and Central Australians, in tolerance to dry heat.

Experiments in humid heat tolerance have been reported on 168 subjects of eleven different groups around the world. A standard four-hour work load was imposed in a portable tent. The main finding was Caucasians living and working in the desert or tropics, once they were acclimatized, did as well as natives living in the same place. The only difference was that white Australians sweated more than aborigines. Unacclimatized Caucasians performed worst of all in humid heat. All of the various indigenous groups maintained similar body temperature, but responded differently in other respects. The desert-dwelling Arabs, for example, sweated more than tropical Bantu or Australian aborigines. As to pulse rate, a low pulse indicated efficient performance. Bushmen performed the given work load with the lowest pulse rates, below those of Bantu, desert Arabs, or Australian aborigines. These results must be treated with some reserve, since the samples were small and the acclimatized Caucasians highly selected. The general run of acclimatized Caucasians might have performed quite differently. Here again, the lack of population differences show that man is a tropical animal, and that adaptation to heat is an ability shared by all men.

Nose Shape. It was once thought that the shape of the nose varied with latitude, that is, simply with heat; it now appears that humidity is the key, through the ability of the nasal membranes to moisten inspired air. People in the moist tropics have short, broad noses, like Africans and Southeast Asians; those in the dry tropics have long, narrow noses, like Arabs or East African Hamites. Eskimos in the dry Arctic likewise have long, narrow noses. Like most generalizations, we find exceptions here as well: both the Bushmen and the Australian aborigines, who live in tropical deserts, have broad noses. So do many Mongoloids.

Altitude

Studies of adaptations to high altitude have been most extensively carried out in the high Andes, but other less comprehensive work has been undertaken at Leadville, Colorado, among Sherpas and Tibetans in Nepal, among Kirghiz tribesmen at seven to nine thousand feet in the Soviet Union, and at similar heights in the Ethiopian highlands. Concentrating on the Andean natives and neglecting some conflicting data from other populations, some slight adaptive changes, compared to lowland dwellers, are detectable even as low as six thousand feet. Life at high altitude imposes a complex ecological stress, consisting of three factors: (1) low barometric pressure, (2) low moisture content of the air, and (3) cold. The low barometric pressure acts by lowering the oxygen tension in the inspired air and the oxygen and carbon dioxide tension in the lung. In addition, mountainous terrain is difficult, requiring high levels of muscular activity.

Adaptation to altitude is both short-term and long-term. The limit for work in the unacclimatized person is about eight to ten thousand feet; at thirteen thousand feet, the rate of oxygen transfer across the lung membranes to the blood is insufficient even for sedentary existence. The body responds to altitude in several ways, chiefly by increasing the production of red blood cells (secondary polycythemia). This increases the oxygen-carrying capacity of the blood. The rate of respiration also increases, as does the depth of respiration. At fifteen thousand feet, acclimatization takes about ten days. In Andean natives, but not in other peoples born and resident at high altitude, one sees a barrel chest and enlarged lungs.

More important than these anatomical changes are the functional or physiologic adaptations to altitude. The Andes people born and residing at high altitudes tend toward the high side of normal in both red cell counts and in their ability to transport the available oxygen to their tissues. Fertility is decreased in some but not all Andean communities. Birth weight is lower and growth is slower, with delayed adolescent growth spurt, sexual development, menarche, and age at which women bear their first child. The placenta enlarges, providing more oxygen to the fetus. Infant mortality is increased at higher altitudes.

Several Andean animals—the llama, vicuna, and the Bolivian goose—native to high altitudes, all have high red blood cell counts and increased affinity of the hemoglobin for oxygen. These characteristics are retained by the first animal generations born at sea level.

The Mongoloid Habitat. The highest mountains in the world are inhabited by Mongoloids: the Tibetans live up to 15,000 feet and the Peruvian Andeans at 17,500 feet, with daily trips to the mines at 19,000 feet. But some members of all major racial groups live in mountains, although to be sure at lower altitudes. These include, among Caucasoids, the inhabitants of the Caucasus Mountains themselves in Russia, the Kashmiri in northern India, at 10,000 feet, and both white and black Americans at Leadville, Colorado, also at 10,000 feet; Negroid people in

Ethiopia live below 11,000 feet and somewhat lower than 7,500 feet in Basutoland. In the Pacific, the New Zealand Maori and New Guinea Highlanders live at moderate elevations. Thus all human groups are capable of some degree of altitude acclimatization. Whether all could adapt to extreme altitudes as well as Mongoloid populations is not known, but perhaps the critical altitude for such adaptation in any human group is about 10,000 feet.

Incidentally, the Mongoloids seem to live in the most diverse habitats of any major group of mankind: from the Arctic and Tierra del Fuego to the moist jungles of Indonesia and South America, and in the world's highest mountains, in Tibet and South America.

Nutrition

By strictly American standards, over half the world's population has a food intake barely sufficient to maintain life. Yet in terms of fertility, work load, and longevity, this ill-fed half is doing a great deal better than merely hanging on to life. Some human populations have made extraordinary adaptations to diets that would send others to mass graves, even after allowance is made for variables such as body mass and ambient temperature. Gross indications of such adaptations to under-nutrition can be seen in the slow growth and late maturation of children, resulting in small size of adults. These are short-term, individual adaptations, which are immediately reversed when children are given adequate nutrition. If they were genetic adaptations, they would not reverse so quickly.

Effects of Malnutrition. A question causing considerable controversy is the extent to which bodily and mental growth are permanently impaired by early malnutrition. My conclusion, from sifting the evidence and conflicting claims, is that the earlier, more severe, and longer-lasting the malnutrition, the longer does the deficit last. However, physical growth and development usually returns to normal, illustrating the principle of catch-up growth. Incidentally, the importance of reaching full growth can be seen by some statistics on type of delivery and infant deaths in relation to height. In one study in Scotland, there were ten times as many Caesarian births and four times as many infant deaths, due to birth trauma, in women under five feet tall as in women two to four inches over five feet.

Mental Effects. With regard to mental development, early malnutrition severe enough to send a child to the hospital does have a long-term effect. It decreases measured intelligence as well as the ability to learn basic academic skills. But I should also point out that malnutrition of this degree is extremely rare in the United States. From the human point of view, some areas of the world are distinctly deficient, nutritionally. The tropics represent both the small end of the Bergmann's-rule distribution of body size (according to temperature) and the hypocaloric or under-nourished part of the world. Many vitamins and minerals are absent in the tropics; parasites are abundant and further reduce the available energy. Other nutritionally deficient areas are middle America and the Andean highlands.

Racial or Population Differences; Lactose Intolerance. Contrariwise, human dietary needs and tolerances vary widely the world over. Many groups—in Ceylon, in Peru, and among the Bantu, for example—have very low calcium intake, but seem to do quite well with regard to work and show no calcium-deficiency diseases (like rickets). Other groups deficient in vitamin C, like the Bantu, nevertheless show no scurvy. The reason for such differences may be in part genetic, since those who fail to adapt to nutritional deficiency will die. A real genetic difference is the deficiency of the enzyme lactose among most Negroid and Mongoloid adults. This prevents them from digesting the milk sugar, lactose. Obviously, all infants can handle milk, since this is what they get from their mothers. But in most populations in the world, outside of northwest Europeans, adults and a goodly proportion of children have lost the ability to digest lactose. When they drink milk or eat ice cream, they experience bloating, diarrhea, and abdominal pains. This enzyme deficiency doubtless acounts for the failure of some feeding programs, both at home and abroad, that depend on milk products and obviously has important implications for our war on hunger. Fortunately, there is a simple solution: ferment or acidify the milk. Sour cream, yoghurt, and cheese have much or all of the lactose removed, while retaining all the nutritional value of whole milk—specifically, protein, fat, Vitamin D, and calcium.

The high infant mortality of many populations due to marasmus (combined protein-calorie deficiency), kwashiorkor (protein deficiency), or vitamin deficiencies like beriberi, imposes a stringent selection. In Guatemala, for example, at least 50 percent of all deaths between one and four years of age are due to malnutrition. The survivors of such stringent selection must indeed be increasingly adapted.

In industrialized countries, overnutrition is a real problem. Obesity is a risk factor in the development of diabetes, hypertension, and coronary heart disease. Persons who are more than 20 percent overweight for their age height and build have a shorter life expectancy than those of normal weight. Our national diet, which is high in fat and in total calories, combined with reduced physical exertion, places a premium on the individual's ability to handle the combination without developing atherosclerosis. On the other hand, these are diseases of later life, so that selection against them is not as stringent as for conditions which occur before or during the reproductive years.

Genetic Plasticity

In surveying human adaptation to environmental stress, we have been discussing traits that are largely continuous variables: anatomical and physiological features whose genetic background we do not know. However, things like body shape or resistance to cold are surely polygenic, not single-gene traits, and this means that they are more likely to have a plasticity response both for the individual and for the population. In the latter case, response to environmental stress can take place more smoothly and rapidly from the genetic resources of the population. In the individual's life span, responses such as slow growth and maturation of undernourished children may be based on a genetically determined plasticity, that is, under the control of genes that in fact affect the action of other genes. These gene complexes may vary from one population to another, although we know nothing about them factually. We have only observed the association between environment and some phenotypic features which exist and which show that small-scale evolution in man is a reality.

While certain groups may not do well in certain habitats, for example Caucasians in the moist tropics, or Africans in extreme cold and possibly extreme altitude, it is still clear that the adaptive range of modern man is tremendous. Much or most of man's adaptive capacity, however, is cultural, and as man increasingly creates his own environment, special physiological and morphological adjustments will become less and less important. This does not mean that man will not be under pressure to adapt, as we shall see in the next chapter.

SUGGESTED READINGS

Damon, Albert. 1975. *Physiological anthropology.* New York: Oxford University Press.
Dubos, R. 1967. *Man adapting.* New Haven: Yale University Press.

15
THE MAN-MADE
ENVIRONMENT

We have surveyed forms of human variation which seem to be adaptations to the world around us, particularly to the weather. Put coarsely, this is a view of an animal undergoing small-scale evolution, bringing about population differences, over many thousands of years.

The Role of Culture

This treatment ignored culture, without denying its presence. Culture, at least in the form of manufactured tools, is as old as the genus *Homo*, about two million years (Figure 15.1). It is the factor which eventually allowed man to break out of the tropics, and to occupy those very regions we have considered as calling for some special adaptation: the arctic, very high lands, and deserts (Figure 15.2). Providing he can find food, man can put up with fairly severe climates, as in southernmost South America, with little clothing or shelter, although even in this case the Indians of Tierra del Fuego like the Alacaluf seem to have adapted by means of thick limbs and lateral body build. But in the arctic, or wherever there is real continuous frost, life is unthinkable without proper clothing and housing. And even in hot, sunny deserts, although black Africans of the Sudan can go naked, light-skinned Arabs could hardly survive without being well covered; and neither people, of course, could manage without being able to drive four-footed food long with them.

Thus culture made possible the expansion of man into some of those

15.1 Man's development and use of tools—actually extensions of his limbs—are what distinguishes man from other animals. While perhaps a million years separate the crude chopping tool (a) from the elegant handaxes (b) and (c) (tools which survive from the lower Palaeolithic), man's technological development is one of amazing acceleration. Man has learned more about flying in the past seventy years than the bird has in seventy million.

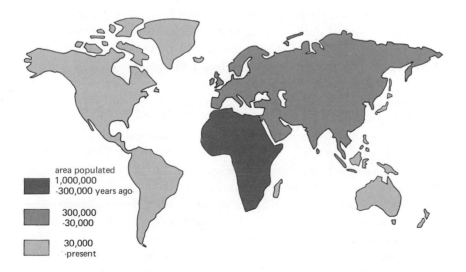

area populated
1,000,000
-300,000 years ago

300,000
-30,000

30,000
-present

very regions where the particular stresses of the last chapter could play upon his physique and physiology. At the same time the protection culture affords has made it a factor for the unity of mankind. Another kind of mammal, becoming isolated and adapted to different habitats, would eventually form two or more different species, but not man. Other cultural features act to maintain contact between populations. One is the incest taboo, which encourages outmarriage and thus social ties among local groups. Another is language, which can serve to isolate speakers of different languages, but, nonetheless, is a kind of communication available to all mankind. So, while major races are the approximate equivalent of subspecies in other animals, we remain a single species, *Homo sapiens*.

This trend to isolation and to local diversity, both physical and cultural, was strongest when technology was simple. But during the last three hundred generations, or roughly ten thousand years, four technological revolutions have taken place, earlier in some societies than in others. These were, in turn, agriculture, city life, advanced technology, and industrialization (Figures 15.3 and 15.4). As a result, the protective screen of culture, interposed between man and nature, has become highly effective for some populations, and is on its way to doing the same for all (if other disasters, like overpopulation, do not prevent it). We must always remember, however, that by far the greater part of man's biological existence, as *Homo erectus* and then as *Homo sapiens*, took place before these revolutions.

The New Environment. Technology today is highly efficient. As a result, one of the most striking aspects of human biology today is that man now virtually creates his environment. Eskimos have oil or gas heat. They wear mass-produced clothing made in the industrialized countries. They drive snowmobiles instead of dog sleds, and they eat fish from cans, shipped from Japan, rather than fish that they themselves have caught through the ice. People living in the tropics now have electric fans, wear protective clothing, and use tinted sunglasses. Soon we may expect them to live and work in air-conditioned quarters. It is this new, man-made, artificial environment—one which can change rapidly and radically, in unpredictable ways—to which man must adapt in the future (Figure 15.5).

What are the components of such an environment? They are seen in most advanced form, of course, in the developed, or industrialized countries, but they began to take shape some nine or ten thousand years ago, when the invention of settled agriculture and the domestication of animals freed men from their natural existence as hunters and gatherers. The steady source of food made it possible for populations to grow larger. For greater efficiency in farming, and for protection, people moved together

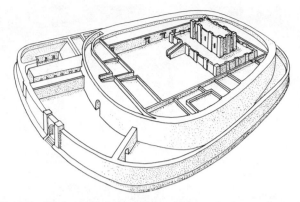

15.3   The Temple Oval at Khafaje. The invention of agriculture in the Mideast ten thousand years ago permitted man to develop monumental architecture, religious and government institutions. (Redrawn from Delougaz, P. 1942. *The Temple oval at Khafaje.* Chicago: Oriental Institute Publication, 58.)

into villages, some of which grew into cities. Large concentrations of people increased the spread of disease and increased stimulation of all kinds—sensory, social, and intellectual. Specialization and competition increased; at the same time, physical exertion decreased.

Modern technology, increasing exponentially with the development of science and its conscious application, intensifies all of these problems. Cities become larger, noisier, and more crowded. Socially defined goals, with delayed rewards, replace straightforward work, which produces goods or services for immediate satisfaction. Travel to work or for recreation becomes increasingly difficult. Far more time is spent in states of arousal, in the need for constant decision, in potential conflict, and in goal-directed activity than in the simpler rhythms or cycles of work-eat-play-rest of early man, dictated by the reasons and by the daily round of

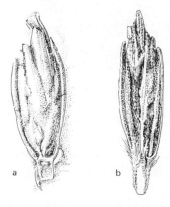

a   b

15.4   (a) Imprint of wheat accidentally left upon clay some eight to nine thousand years ago at Jarmo, an archaeological site in northern Iraq. Archaeologists interested in the beginnings of agriculture compare these with modern varieties (b). (Redrawn from photographs in Braidwood, R. J. 1960. The agricultural revolution. *Scientific American.* 203:136.)

15.5 Toward a global culture: A Guinness beer truck making deliveries in Singapore. (courtesy Peter Timms)

day and night. A new set of diseases, the so-called diseases of civilization, replace the infections and parasites of early man. Although health and longevity have improved, they depend on a precarious balance of social stability and continuous monitoring of the environment.

Now this is the environment of modern man, toward which all of mankind is moving. Urbanization and industrialization are proceeding at a rapid pace all over the globe. What are the components of this man-made environment, and how do they affect man biologically? These stresses are on the whole subtle and long-lasting. Some of them are physical, some are biological, and some are psychosocial. They include artificial illumination, noise, crowding, disease, constant alertness and arousal, conflict, insecurity or even fear, mobility, and constant and accelerating change. One concept that summarizes many features of the modern environment is "information input overload"—which might be vulgarly paraphrased as "just one damn thing after another"—or more accurately, "just too bloody much." Experiments have shown that information input overload is harmful to many forms of life, from unicellular organisms to man, and even to machines. Telephone systems, electric power grids, and computers break down when the circuits are overloaded. Quite apart from the total number of possible stimuli, many of ours are attractive, which makes them even harder to avoid. Our environment also contains chemical and physical mutagens, many kinds of toxins or poisons, and widespread pathogenic microorganisms. As mentioned, our environments disrupt our age-old biological rhythms.

We obviously cannot study all of these factors in detail. The trouble is that very little is known about their more subtle, long-term effects on man. Despite the alarms raised, the hands wrung, and the dire predictions of doom based on animal experiments and observations, very little research has been done on the effects of light, noise, and crowding on man himself. Let us have a look at what is known.

Light

First of all, let us consider light. In the last chapter I mentioned the effects of ultraviolet light reaching the earth from the sun. Ultraviolet light causes sunburn, tanning, and skin cancer, and promotes the formation of vitamin D in the skin. In turn, vitamin D increases the absorption of calcium from food, and thereby promotes bone growth. Sunlight also destroys the elastic tissue of the skin, causing wrinkling in old persons. The elastic tissue of unexposed skin is perfectly normal, even in old age. So it is sunlight, and not aging, that is responsible for the worst manifestations of senile skin. Heavily pigmented people, whose melanin screens out sunlight, have much less wrinkling of their exposed skin than fair-skinned persons.

Very recently, another direct biological effect of light has been discovered: it can degrade, or break down bilirubin, a yellowish pigment formed from hemoglobin when red blood cells are destroyed. This pigment gives the yellow color or jaundice to the skin of people with liver damage. Now some infants are born with jaundice, mostly from Rh incompatibility; as you will recall, an Rh^+ fetus born to an Rh^- mother may show this condition. Treatment with light will dissolve the bilirubin, clear up the jaundice, and cure the infant until his own red cell production can catch up with his needs. This discovery raises the possibility that light may have other major biochemical effects, as yet unsuspected. Like X rays, light may be beneficial in some situations, but may already be impairing human health by destroying essential compounds or generating toxic ones.

New Artificial Lights. Americans spend much of their time indoors, exposed to artificial lights whose spectra differ considerably from sunlight. Incandescent bulbs, like the sun, are radiant, heated blackbodies, and their spectra resemble sunlight. But fluorescent light works in a very different way, not by heating photons, but by exciting chemical phosphors by a stream of electrons. Fluorescent lights are engineered for visual brightness and therefore emit more yellow and less red light than the sun. It is conceivable that this unplanned exposure might have physiological consequences. Although mankind has far more urgent problems, it is certainly necessary to find out more about the biological effects of this completely new environmental agent.

Light and Daily Rhythms. Light has two kinds of effects, the direct ones we have just mentioned, and the indirect ones, mediated by light receptors in the retina of the eye. Vision, of course, is the most important of these indirect effects, but there is a large and expanding list of neural and glandular effects. These include control of sexual maturation, ovulation, and a large number of daily rhythms. Rather unexpectedly, blind girls show an earlier menarche than girls who can see. There are different rhythms of growth in blind children, throughout the years, compared with those who can see. The mechanism whereby light influences sexual maturation is not entirely known, but the pineal gland, in the brain, plays a large part. (Descartes thought the pineal gland was the seat of the soul but it seems to be just another gland which is somewhat more complex in activity than some others.) The daily cycle of light and darkness generates, or in the past initially determined, many circadian or twenty-four-hour cycles, such as body temperature, sleep, and production of many hormones and enzymes. Plants and animals also have such cycles. Some animals have annual cycles, like the migration and mating of birds, or the testicular maturation and the ovulation of sheep, or hibernation in many animals, all depending on the length of the day. Whether man has an annual cycle is unknown, although a few observations suggest that he may. Nothing is known about the wavelengths of light responsible for these events. Most biological rhythms originally dependent on light have become so well-established that they persist even when the animal or the person is completely shielded from light.

Other biological rhythms are much more difficult to explain, such as the rat's four- to five-day period for ovulation, or the human menstrual period of twenty-eight days or so. Presumably the lunar cycle was originally responsible, but it is hard to imagine how this effect became established.

Incidentally, it has been reported that Eskimo women do not menstruate or deliver children during the long, dark winter. This is quite untrue. A Harvard anthropology student studied church and medical records of a Labrador Eskimo village founded two hundred years ago, and found births had taken place in all months of the year, although with a strong seasonal pattern. Even in New York City there is a seasonal pattern of births: most occur in the summer, having been conceived in the late fall.

So in summary, light is an important aspect of the environment, biologically. Man has now introduced an entirely new factor, fluorescent light, to which large numbers of people are exposed for long periods of time. Modern life disrupts long-established biological rhythms based originally on the daily cycle of light and darkness. We do not know what all this is doing to us, and we had better find out.

Noise

Nobody in a modern city needs to be told that we live in a noisy environment. Try lying down and counting the number of minutes during the day that you cannot hear an airplane. If you live west of Boston in the air corridor to New York City, you will not find much time free of the sound of airplanes. And this of course is only one source of noise. All machinery is noisy, from household appliances to automobiles and earth movers. Lawnmowers, snowmobiles, and unsolicited motivational or mood music in stores, elevators, airports, and airplanes are other horrors we have devised for ourselves, topped by rock and roll music.

Effects on Hearing. The direct effect of all this noise on the ear has been thoroughly documented. Urban, industrial populations lose much more hearing with age, and have worse hearing at all ages, than rural or preindustrial populations. Factory workers, aircraft mechanics, or construction workers have much worse hearing than persons who work in quieter surroundings. College freshmen, during the peak years for rock and roll music, had the amount of hearing loss, particularly for high tones, usually found in persons aged sixty and above. After only two hours of exposure to sound levels in discotheques, sixteen percent of young people were found to be adversely and possible permanently affected. The wide publicity given to these findings may, one hopes, reduce the decibel level in discotheques. (Noise is measured in decibels, a relative scale based on ratios of noise to a standard whisper. Man can distinguish 130 steps, at which point noise becomes pain.)

Side Effects of Noise. In the above paragraph, we have been referring to loud noise. The effect of moderate noise levels is not so clear-cut. Interference with sleep produces inefficiency and irritability. While prolonged noise has no effect on the performance of routine work, skilled jobs which require concentration suffer a loss in efficiency, which can run as high as 70 percent.

Although people claim to be annoyed by moderate noise, there is little objective evidence of impaired health or behavior. In one series of experiments, Glass found that the kind of noise that impaired performance most was unpredictable and uncontrollable noise; people can adapt to steady or predictable noise. Of course, the possible cost of such adaptation, which is not measured in the experiment, must be considered. The perfect example of a sudden, loud, unpredictable, and uncontrollable noise is the sonic boom of the SST. This would account for the violent annoyance of people subjected to sonic boom.

A fascinating footnote to this discussion: Over a two-year period, persons living in the maximum noise area around the London airport had a higher rate of admission to mental hospitals than persons living outside that area. In research that I (with students) have conducted in a large housing project in Boston (Figure 15.6), we found that people living in a noisy area have higher rates of arrests and school dropouts than those living in a quieter area. Maintenance of the dwellings, for which the tenants are responsible, was much worse in the noisy area. We realize that these associations tell us nothing about cause; it is possible that problem families are assigned to the noisy area. We have investigated this possibility and find only that the noisy area does indeed have more large families, since there happen to be more large apartments in those buildings. Large families may be problem families by virtue of their size, but this seems unlikely to account for all of the difference in their behavior. The answer will come in a few years. The bridge whose traffic causes the noise is being

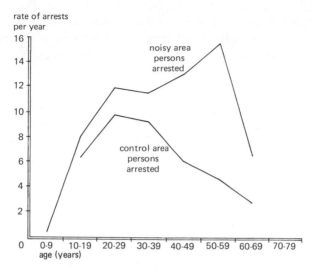

15.6 "Jamesville" housing project, arrest rates by age, 1969, 1970. A study of more than one thousand individuals in a pseudonymous housing district of Boston. Approximately one-half of the individuals lived in apartments in which outside noise levels were high. The others, a control group, were in a quieter area of the same project. (Data were compiled by the author.)

designed to reduce the horrendous noise levels caused by the constant passage of huge trucks, day and night. The noise reaches a continuous level of eighty to ninety decibels, which is known to cause hearing loss when continued over an eight-hour working day in factories. There are occasional peaks up to one hundred and ten decibels, which is the noise made by a motorcycle. When noise levels are eventually reduced, it will be instructive to look at the new arrest and school dropout rates of the same families living in the same houses, but in a much quieter environment.

Difficulties of Noise Control. What can be done to lower noise levels? This is much harder than one might think. Clearly we should reduce or eliminate all unnecessary noise; my own favorite candidate for elimination is the mood music in airports, airplanes, hotels, and supermarkets. But probably most people really like mood music—not only the supermarket owner. A certain amount of noise is unavoidable in an industrial society. The cost of reducing noise must be balanced against the cost of retaining it. In industry, for example, cutting the noise level in half will reduce it by only three decibels because the decibel scale is logarithmic. Say you have one hundred machines, with a noise level of one hundred decibels. You cut the noise source in half by stopping fifty machines, and the noise level remains at ninety-seven decibels. You have reduced production by

50 percent and thrown 50 percent of the workers out of their jobs, for a negligible decrease in noise. Furthermore, noiseless machines are expensive, adding to the cost of the product. So things are not as simple as they seem. In fact, some businesses solve the problem of noisy machines by hiring deaf workers!

This is our introduction to the notion of the trade-off. This can be stated in another form: you don't get something for nothing. Everything you do, and everything you don't do, has a cost. There is no free lunch. We shall meet this principle many times in the rest of the book.

Crowding

No aspect of modern life is more conspicuous, or has received more attention, than crowding. We are all painfully aware of population increase. The animal evidence is spectacular. What does crowding do to man?

First, let us define crowding. If we look at density, which is that number of people per area or per square mile, we find that countries vary widely. India has five hundred persons per square mile, Japan has seven hundred, Holland has eight hundred, and the United States has only fifty-five. But localized densities are much higher: Tokyo has twenty thousand, New York City has twenty-five thousand, Manhattan has seventy-five to one hundred thousand per square mile. Downtown on a weekday, Manhattan has a density of three to four hundred thousand per square mile (Figure 15.7).

To anticipate, microcrowding, or density per area of living quarters, is much more important than macrocrowding, or density per large unit of area, such as a city district or block. A prison cell has forty square feet for one person, of which the bed occupies twenty. Military barracks give thirty square feet per person, troopships (with quadruple bunks) twelve square feet, a New York night club ten, and a theater at intermission, eight. The Black Hole of Calcutta or the New York subway at rush hour provide two square feet per person.

One must also distinguish crowding from overpopulation. Overpopulation implies insufficient food, overconsumption of resources, pollution, dirt, noise, heat, odors, discomfort, and the like. Crowding, which is simply a high ratio of people to area (Figure 15.8), tends to increase these extraneous sources of discomfort. If you could engineer them out, what would be the residual effect of crowding itself on behavior, performance, and attitude?

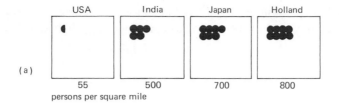

15.7 (a) Population densities in selected countries. (b) City crowding in Tokyo and New York. Note that in the first figure each dot represents one hundred people; in the second, each represents ten thousand.

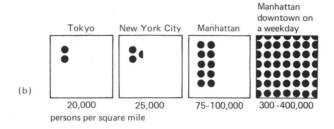

15.8 Man crowds the landscape. A Lamasary, Labrang, Tibet. (courtesy Peabody Museum, Harvard University)

Effects on Animals. The animal evidence comes from observations by Konrad Lorenz and others on territoriality, showing that animals insist on a certain amount of space which they defend against intruders of their own species. Territoriality is a response to population density and, together with dominance, provides a mechanism for controlling population density. Famous experiments by Calhoun on rats and by Christian on deer show what happens when these animals experience uncontrolled population growth in a limited area. There are three main kinds of effects: (1) a breakdown of reproductive behavior—mating, nest building, delivery, and suckling; (2) a high frequency of abortions, stillbirths, maternal and infant deaths; and (3) increased aggression, particularly by males. Some animals became abnormally passive and withdrawn. These were all short-term responses. Studies of long-term adaptation to crowding show a somewhat different picture. Kessler found that under extreme conditions of crowding in mice, there was no increase in pathology once the population had achieved its maximum density, and no further population growth was occurring. Under these circumstances asocial or antisocial behavior was common, but there was no more disease or mortality than in a control group living under uncrowded conditions. In other words, mice adapt to a high density of population, just as people do, if the population growth is slow enough. In other experiments on animals, the effects of crowding on disease have been contradictory. Solitary animals are more subject to disease; and sometimes overcrowded ones show the same effect, sometimes not.

Effects on Man. Despite the inadequate or contradictory evidence from animals, the dramatic findings of Calhoun and Christian have been extrapolated to man. There has been much breast-beating and prophecy of gloom and doom. But people do not behave like animals. Clearly population density cannot continue indefinitely without affecting health and behavior, but even the maximum densities now existing have no apparent ill effects. As to evidence from actual research, very little work has been done on human response to crowding. Of three studies which tried to relate population density of city districts to disease and crime rates, in Honolulu, Chicago, and other American cities, two studies showed no effect of density. In a few other experiments on small groups of people, behavior was more aggressive, increasing with more people and with higher density of people per room. There is also a body of anthropological observation showing that various cultures define and use space differently, some preferring close contact and others the opposite. Eastern and Southern Europeans interact at close range, Northwest Europeans and Americans at a distance. Densely crowded countries as distant as Britain and Japan have developed some striking similarities in the way they handle the problem: a formalized class structure in which everyone knows his own place and that of others, and a fanatic regard for privacy and private gardens. The formal class structure makes it easy to handle contacts with large numbers of persons, since one does not have to establish appropriate responses afresh with each new person—your response is automatic.

Effects on Social Behavior. As for the effect of crowding on human performance and attitude, virtually the only experiments so far have been conducted at Stanford by Freedman and Ehrlich. These investigators placed their subjects for four hours in a very crowded room, with chairs touching, allowing four square feet per person, and again, for another four hours in a less dense room which allowed fifteen to twenty square feet per person. In both rooms, chairs were comfortable, temperature was controlled, and air conditioning removed odors. So the crowding was intense for the four-hour span. The results were as follows: in a number of intellectual tasks requiring a great range of intellectual ability, there were no effects of any kind attributable to the crowding in the rooms. Males and females behaved alike, as did younger and older subjects. In contrast, crowding did affect interpersonal behavior. The degree of crowding influenced competitiveness and severity of sentencing (in a mock trial). In tests of the sexes conducted separately, men became more aggressive, competitive, and hostile with crowding, but the women less so! Women liked the high density, becoming more friendly. When men and women were mixed, these opposite trends canceled out, and there were no effects of density.

The Unaffected Lau. In the Solomon Islands, one of the tribes we studied was the Lau. These tribesmen live on small artificial islets in a shallow, salt-water lagoon on the edge of Malaita Island (Figure 15.9). The artificial islets, some of which are hundreds of years old, are made by transporting soil and coral from the main island, and are less than one acre in area, but are inhabited by two hundred to two hundred and fifty people. The density is eighty thousand people per square mile, a figure reached in parts of Hong Kong, Calcutta, and Johannesburg.

The striking finding is this. Despite a density as high as that reported for any human population and despite some minor infections, the Lau enjoy robust physical and mental health.

This observation bears directly on environmental and urban design, as well as on our thinking about human ecology. It might be possible to design healthful residential environments for much higher densities of people than we now contemplate, if we knew what protects the Lau from the disorders often attributed to overcrowding among animals and urbanized man. The answer is certainly not simple; many factors are at work. The Lau have a fairly small absolute number of daily encounters, all

15.9   Lau Lagoon islet, Solomon Islands. (courtesy W. W. Howells)

with kin or close acquaintances. Spatially, each islet is divided into three areas: one reserved for men, one for women, and one a common portion for both sexes, containing family dwellings. There is easy access by canoe to unlimited, unpeopled space on the surrounding sea or on the mainland, a few hundred yards away. New islets are continually being constructed to relieve overcrowding. Adults typically leave the crowded islets during the day, the men to fish and the women to garden and fetch water. That is, time is traded for space. Even on the islets themselves the same holds, with most of the men's time and much of the women's time (all of it, during menses and childbirth) spent in the uncrowded areas reserved for either sex. And finally, Lau society is technologically simple, lacking many of the sources of tension found in industrialized settings and perhaps aggravated by crowding.

Such possibilities suggest hypotheses for further test among appropriate human groups, each a unique natural experiment. However intriguing the analogies from animal behavior, only research on man can solve mankind's most urgent problem, learning to live in the human ecosystem.

SUGGESTED READINGS

Banfield, Edward C. 1970. *The unheavenly city, the nature and future of our urban crisis*. Boston: Little, Brown.

Bernarde, Melvin A. 1970. *Our precarious habitat.* New York: Norton.

Ehrlich, Paul, and Ehrlich, Anne. 1972. *Population, resources, environment,* 2d ed. San Francisco: Freeman.

Oakley, Kenneth P. 1960. *Man the toolmaker.* Chicago: Phoenix Books, University of Chicago Press.

Vayda, Andrew P. 1969. *Environment and cultural behavior.* New York: American Museum of Natural History, The Natural History Press.

Wagner, Richard H. 1974. *Environment and man,* 2d ed. New York: Norton.

16
RADIATION IN THE ENVIRONMENT
By Donald E. Gerson

Types and Effects of Radiation

Our environment is constantly exposed to natural radiation from the sun and stars. The effect of this radiation depends on its energy, which is variable and is related to its wavelength. The very high energy radiation includes cosmic and gamma radiation. At somewhat greater wavelengths and lower energy are ultraviolet radiations; infrared radiation is of even greater wavelengths. Radio and electrical waves have the greatest wavelength and lowest energy.

The effects of a given radiation depend on its wavelength and energy. A thin band of radiation between ultraviolet and infrared is capable of stimulating receptors in the human eye and constitutes visible light. Somewhat higher energy ultraviolet radiation is capable of producing tanning or sunburning in human skin. The much lower wavelength gamma and cosmic radiations are of such high energy that they are capable of breaking chemical bonds in molecules and of passing through solid objects. This latter quality is utilized in one form of man-made radiation (a type of gamma radiation) referred to as X rays. This useful quality of high energy radiation, however, is somewhat offset by its ability to sever chemical bonds in molecules. Such radiation can alter or destroy biological compounds and thus potentially do harm to living organisms.

A vast amount of radiation bombarding the earth is filtered out by the earth's atmosphere and so does not affect man. A small amount does reach the earth's surface and constitutes part of the natural background radiation to which we are all daily exposed. Perhaps far more important than this, however, is the high energy radiation produced by man. This includes medical radiation, radioactive fallout from thermonuclear bombs, and radioactive waste material from nuclear reactors. The exposure to man from these has tended to increase in modern times and probably will continue to increase in the future.

Harmful Effects of High Energy Radiation

Mechanisms of Action. As mentioned above, high energy (gamma and cosmic) radiation is capable of breaking chemical bonds in molecules. Any molecule in a given organism can be so affected, but the most significant effect occurs when the molecule involved is DNA. This is due to the central role of DNA in the replication of cells and in the storage of information for the production of the specific porteins and enzymes which a given cell requires to carry out a specific function. Thus, if a molecule of DNA is destroyed or altered by radiation, the cell containing this molecule of DNA may be affected in one of several ways:

1. The DNA may be sufficiently damaged that the cell can no longer carry out its functions and dies (often during the process of replication).

2. The cell may survive but its DNA may be altered in some subtle way which causes the cell to lose the ability to control its own rate of replication. Such a cell may begin to divide rapidly and uncontrollably and will then constitute a neoplasm, or cancer.

3. The altered molecule of DNA may reside in a reproductive cell (ovum or sperm). In this case, if the cell survives, its abnormal DNA may be passed on to a given offspring, nearly all of whose cells will contain a replica of the abnormal DNA, which can then be passed on to his offspring. In short, a mutation will have occurred. Depending on the nature and degree of alteration of the DNA, the mutation may or may not produce recognizable phenotypic abnormalities.

The first two of the above are referred to as somatic effects of radiation and are biologically important only for the individual affected. The last is a genetic (heritable) effect and is potentially of far more importance to the species.

Relationship of Radiation Effect to Dose. Experimentation with animals as well as observation of human beings exposed to radiation from thermonuclear bombs and medical irradiation have led to the establishment of lethal and permissible doses for man. The unit of exposure used is the rem or millirem (mrem). (See Table 16.1 below.) The lethal dose for man, when the radiation is given as a single, whole-body dose, has been empirically established and is given as the LD_{50}—that exposure which would kill 50 percent of an average exposed population. The maximum permissible (safe) dose to an individual is not so readily determined. Lower doses of radiation may produce mutation or cancer, both of which may require many years to become apparent. In addition, at lower exposure levels, such abnormalities tend to occur in only a small percentage of people, even though their ultimate effect on a population or species may be important. Thus, determining whether very small doses of radiation produce significant biological effects requires studying very large populations over long periods of time. Animal experimentation may shorten the length of such a study, but the applicability of the findings to man always remains in doubt.

In short, it is still not clear whether any dose of radiation is safe. For the time being, however, an international commission, extrapolating from human and animal data, has established a permissible annual dose level for man. This is not much above the annual exposure from natural background radiation and is well below the lowest exposure known to date to produce significant biological effects. Table 16.1 gives the actual doses in millirems of whole body exposure for the LD_{50}, maximum permissible dose, natural background radiation, and average exposure from medical radiation.

The biological effect of radiation tends to be cumulative insofar as the damaging effects of any given dose of radiation are superimposed upon damage caused by previous exposure. Thus multiple exposures at doses that are individually nonlethal may ultimately be fatal. The effect is not strictly cumulative. Animal experimentation has shown that a given dose of radiation is biologically less harmful if given over a long period of time than if given as a single exposure, indicating that animal cells are capable of some degree of recovery from radiation injury by reparative mechanisms that are not entirely understood.

Another important variable influencing the effect of radiation is the distribution of the radiation dose. A radiation exposure to the whole body

Table 16.1 RADIATION EXPOSURES

LD_{50} for man (single whole-body dose)	400,000 mrem
ICRP maximum permissible dose to an individual	500 mrem/yr
Exposure from natural background radiation	120 mrem/yr
Exposure due to medical radiology	70 mrem/yr

that would normally prove fatal will usually not be lethal if given to a single extremity. By the same token, any given dose of radiation is more likely to produce mutations in offspring if the radiation exposure is to the area of the gonads than if the exposure is to the head or chest.

Clearly these factors influencing the effect of radiation exposure must be considered in any attempt to assess the threat to man of various types of radiation in the environment.

Somatic Effects of Radiation. The somatic biologic effects of radiation (effects occurring in exposed individuals, as opposed to genetic effects) may be placed in three catagories: early effects, late effects, and effects on the exposed fetus.

The early, or acute, effects of radiation are those that manifest themselves within days to weeks of the exposure and are due to killing of cells and resulting organ dysfunction. The most vulnerable organs are those whose cells normally divide rapidly, since the DNA in these cells must replicate itself rapidly, and any alteration in the DNA will be quickly expressed. In man, these include the blood-forming cells of the bone marrow, the cells lining the gastrointestinal tract, and the cells of the skin. The acute effects of radiation result from damage to these organs and are termed "acute radiation sickness."

Acute radiation sickness begins minutes to hours following exposure to high dosages and is manifest by nausea, vomiting, and diarrhea. This may last as long as four days and is then followed by a latent stage during which these symptoms subside and the individual feels better. The length of this stage varies and corresponds to the time required for depletion of essential cells which cannot be replaced due to radiation damage. This phase may last up to three weeks and will be shorter the higher the dose of radiation. It is almost invariably followed by a stage during which the illness manifests itself in a serious form. At doses on the order of 2000 rem, irreparable damage to the cells lining the gastrointestinal tract will lead to internal hemorrhaging, infection, and death, usually within two weeks. At somewhat lower doses, the gastrointestinal tract may recover only to be followed by failure of the blood-forming cells of the marrow, resulting in anemia, depletion of white blood cells and platelets, and death from infection or hemorrhage within four to five weeks. As noted above, the skin is also sensitive to radiation and often becomes inflamed and ulcerated during the illness, leading to another source of infection. At somewhat lower doses, with intensive medical care, the exposed individual may survive acute radiation sickness, with complete recovery requiring approximately two months. At very high doses (on the order of 10,000 rem), progressive swelling of the brain occurs, and death may ensue within hours of the initial exposure.

The late effects of radiation are those that are manifested only months or years after exposure. The mechanisms of their occurrence are not completely understood, but they have been observed in both experimental animals and in human beings accidentally exposed to large, but nonfatal, doses of high-energy radiation. These include early formation of cataracts (scarring of the cornea of the eye), increased incidence of various types of cancer, and accelerated aging and mortality.

Early formation of cataracts has been observed in animals experimentally exposed to radiation and in human beings whose eyes have been accidentally exposed. Although this may not appear to be a major harmful effect of radiation, scarring of the cornea of the eye with resultant loss of its transparancy may lead to severe impairment of vision.

An increased incidence of various types of cancer following radiation exposure has been repeatedly observed in animals and man. Studies of patients receiving high-dose radiation therapy and radiologists in the past (before modern radiation protection) have shown leukemia to occur more commonly in these groups than in the general population. Children who received radiation to the neck in a certain form of treatment used in the past developed cancer of the thyroid at a higher rate than nontreated children. An increased incidence of skin cancer in irradiated areas of the skin has been shown to occur ten to thirty years after significant exposure, and an increased tendency to develop tumors of the lung has been demonstrated in miners of uranium (a radioactive substance).

Accelerated aging and mortality are also late somatic effects of radiation. Premature aging and decreased longevity have been observed in several species of animals given whole body irradiation. In addition, radiologists in the past, when compared to other doctors, showed an increased mortality, both from cancer and from causes other than cancer. (With the development of modern radiation protection, these differences have not been demonstrated in recent studies.)

When a pregnant woman is exposed to high energy radiation, especially in the abdominal or pelvic region, the fetus is exposed to some portion of the radiation dosage. The effect on the fetus constitutes a form of somatic effect of radiation. If the exposure to a significant dose occurs between conception and the subsequent two weeks (a very susceptible period for the embryo) death usually occurs early in intrauterine development, with resultant spontaneous abortion. If the exposure occurs between the second and sixth weeks after conception—the period of formation of the major organs—an increased incidence of developmental malformations (particularly of the brain, eye, and skeletal system) as well as an increased incidence of neonatal death can be expected. Radiation exposures occurring later in intrauterine life tend to produce late somatic effects, such as an increased incidence of childhood leukemia, rather than increased infant mortality.

Genetic Effects of Radiation. The genetic effects of radiation are those occurring in an individual whose parents (or only one parent) were exposed to radiation before his conception. As mentioned above, these occur because of alteration of DNA in reproductive cells of the exposed individual. Such altered DNA may then be passed on to, and replicated in, his offspring. Since many of these mutations apparently do not produce recognizable phenotypic changes (either because they are not functionally significant changes in the DNA or because they are changes involving only one gene of the gene pair and are recessive), only limited information is available concerning genetic effects of radiation in human beings. The data available on this are drawn from a few populations exposed to radiation from thermonuclear explosions and will be discussed in the next section.

The Atomic Bomb: Hiroshima and Nagasaki

The Studies of the Atomic Bomb Casualty Commission. The destructiveness of a thermonuclear bomb is not limited to the immediate destruction caused by the explosion (Figure 16.1). Persons within a certain distance of the bomb (ranging up to several miles) who are not immediately killed by the explosion or the intense heat may be exposed to high levels of direct radiation from the bomb. In addition, radioactive materials from the thermonuclear explosion, propelled high into the atmosphere, may drift to earth far from the immediate area and produce considerable radiation exposure. The inhabitants of Hiroshima and Nagasaki who survived the atomic bomb were exposed to varying doses of radiation, largely dependent on their distance from the site of explosion. These survivors have constituted one of the unfortunate few large human populations in whom the short- and long-term effects of significant radiation exposures can be studied. For many years, the Atomic Bomb Casualty Commission (ABCC) has followed these populations for this purpose. In order to determine the effect of high energy radiation on human beings, the commission has carefully screened these populations for possible genetic, fetal, and early and late somatic effects.

Genetic Effects. In the search for possible genetic effects of radiation,

16.1   Bomb damage at Hiroshima. (courtesy Wide World Photos)

the ABCC has studied multiple possible indices of genetic damage in individuals whose parents were exposed before conceiving. Rate of stillbirth, death rate of newborns, incidence of congenital malformations, weight at birth and at ten months, and childhood mortality have all been monitored. No statistically significant effect has been demonstrated in any of these parameters to date (Figure 16.2). One other rather ingenious method of measuring possible genetic effects—observation of the sex ratio in offspring at birth—was studied (Figure 16.3). The sex ratio was found to shift in the first ten years following the bomb, as theoretically predicted. Thereafter, no significant effects on the sex ratio have been observed.

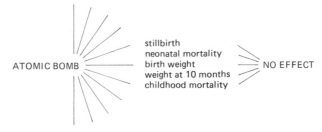

16.2   Hiroshima and Nagasaki. Genetically, there appeared to be little or no effect to those who were conceived after the atomic bomb by parents who had been exposed to it. The single exception appeared to be a decrease in head size (Figure 16.4). (Data from the Atomic Bomb Casualty Commission.)

before radiation

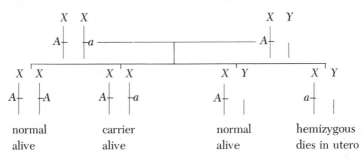

sex ratio at birth: 1♂ : 2♀

sex ratio at birth: 2♂ : 0♀

A = normal gene

a = recessive gene, fatal *in utero* to homozygotes

B = dominant gene, fatal *in utero* to homozygotes or heterozygotes

16.3 Altered sex ratio at birth as an indicator of genetic effect of radiation. Two different groups are considered: one in which only the mother was irradiated, one in which only the father was exposed. In the first situation, if the radiation produced an X-linked recessive mutation in the mother which was fatal *in utero* to homozygotes, only male fetuses could have a fatal genotype. Observing large numbers of such matings, one might expect to see a shift in the sex ratio at birth in favor of females. In the second situation, assume that radiation produced an X-linked dominant mutation in the father which was fatal *in utero* to homozygotes or heterozygotes. Since only females receive the father's X chromosome, one would anticipate a shift in the sex ratio at birth in favor of males.

The explanation for the general failure to demonstrate genetic effects in the offspring of these exposed populations is not entirely clear. One possibility is that the great majority of the mutations produced are autosomal recessives and thus generally do not appear in the first generation of offspring. In this case, an observable genetic effect may occur in the second and later generations, as the first generation offspring who are heterozygous intermarry and produce homozygous offspring. It is also possible that the genetic effect is occurring at a sufficiently low incidence to be statistically undetectable or that the reproductive cells containing mutations do not survive well and are rarely involved in reproduction, and thus the mutations are not passed on to offspring. Perhaps genetic effects will become apparent as these populations are followed in the future and generations beyond the F_1 generation are observed.

Chromosomal Abnormalities. These constitute a special case of genetic effects of radiation, in which DNA is altered in a way that produces a recognizable alteration in the structure of chromosomes. In a sense, they are more easily observed than the vast majority of genetic alterations, in that the aberration is readily observed in any given individual if present. In addition, these changes may be looked for in adults exposed to radiation, fetuses exposed, and in the offspring of exposed parents.

In the Hiroshima and Nagasaki populations, individuals who were under age thirty when exposed, those over age thirty at exposure, those who were *in utero* at time of exposure, and the unexposed offspring of exposed parents were all studied twenty years after the bomb. All groups, when compared to controls, showed a statistically significant, and often large, increase in the number of chromosomal abnormalities, with the exception of the unexposed offspring of exposed parents. Thus chromosomal damage persisting at least twenty years was induced by the radiation exposure, although these chromosomal abnormalities were not passed on to offspring. It is thus debatable whether these changes have any biological functional significance for the species.

Effects on the Embryo. Many of the women exposed to radiation from the atomic bombs in Hiroshima and Nagasaki were pregnant at the time of exposure. As a result, their fetuses were also exposed to varying amounts of radiation. The ABCC has followed those individuals who were exposed to the radiation as fetuses and has observed the incidence of multiple physical and mental anomalies.

Of all the congenital anomalies studied, only one showed any significant effect of radiation. An increased number of children with slightly smaller heads and mild mental retardation was observed (Figure 16.4). This effect was decreased the later in pregnancy the exposure occurred

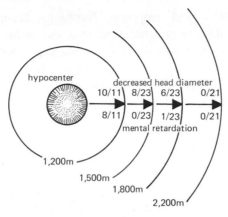

16.4 Hiroshima and Nagasaki. Delayed radiation effects on fetal head size. The fractions above the arrow show the number of cases where the head diameter was more than two standard deviations less than the average. The fractions below the arrow indicate the number of those individuals with mental retardation. While sample size is extremely small, it appears that small head size and retardation is greatest within twelve hundred meters of the blast. Beyond this the proportion of small-headed individuals decreases, and the incidence of retardation is minimal. (Data from Miller, Robert W. 1969, *Science* 166:569–574.)

and the greater the distance of the pregnant mother from the bomb (at distances greater than one mile, the effect was not observed). The apparent dependency of this effect on radiation dose and on stage of development of the fetus suggests that this is a true biological effect of radiation.

Height and weight of survivors who were *in utero* at the time of the radiation exposure were also studied. A statistically significant decrease in height and weight was observed in these individuals, both at adolescence and at age seventeen. The significance of this finding is not clear, since it has been pointed out that the effect may be a socioeconomic effect rather than a biological one. It is reasonable to assume that those individuals nearest the site of the explosion suffered not just the geatest radiation exposure but also the greatest damage to homes and property. Thus economic difficulties may have resulted in poorer care and nutrition for their offspring and thus reduced height and weight (in which case this would not be a direct biological effect of radiation). The difficulty of separating these socioeconomic and biological forces is an excellent example of the problems facing the researcher in human biology.

The ABCC also studied fetal and infant mortality in the instances where the fetus was exposed to radiation from the bomb. Pregnant females who were within 1.25 miles of the explosion were separated into two groups, those who reported symptoms of acute radiation sickness and those who did not report symptoms. The former group presumably received a higher dose of radiation, and fetal and infant mortality was 43 percent in this group, versus 9 percent in the less-exposed groups.

Thus some adverse effects were demonstarted as a result of human fetal exposure to radiation, but these were less common than might have been expected, involved very few anomalies, and tended to be seen at rather high exposures.

Somatic Effects. The survivors of Hiroshima and Nagasaki have been closely examined for possible somatic effects of radiation.

A number of anthropometric measurements were made on children who were ages one to thirteen when exposed to radiation from the bomb. These measurements were made six, seven, and eight years following the exposure. Compared to nonexposed children, this group showed statistically significant, although small, decreases in body measurements at all ages and a decreased growth rate at postpubertal ages. This effect increased with increasing estimated exposure. Once again, this finding is complicated by the possibility that at least part of this effect may be due to economic factors.

The study of the atomic bomb survivors also confirmed the relationship, found elsewhere, of increasing incidence of cataracts following significant radiation exposure. This was detected only in those exposed to rather high doses, and in almost none of the cases were the cataracts severe enough to produce significant visual impairment.

Confirmation of another known sequel of radiation exposure—increased incidence of cancer—has also come from the study of Hiroshima and Nagasaki survivors. An increased incidence of leukemia was found, and this showed a definite relationship to estimated radiation dosage. The peak incidence occurred in 1951, but leukemia continued to occur at higher than normal rates in survivors through 1966. An increased number of cases of thyroid cancer were also detected, and this, too, was found to be proportionate to radiation dose.

Finally, in those survivors who were within three-fourths of a mile of the explosion, mortality from all causes (not including leukemia) was found to be increased by 15 percent. This trend was observed up to 1955, after which the mortality rate in survivors returned to normal.

In summary, then, these studies of the atomic bomb survivors have provided information on the effects of varying large doses of radiation on human beings. Many of the effects theoretically predicted and observed in experimental animals (and small numbers of human beings) have been confirmed in this rather large population: chromosomal abnormalities, certain fetal effects, and several somatic effects, such as mild growth retardation, increased cataract formation, increased incidence of leukemia and thyroid cancer, and elevated mortality rate. On the other hand, many of these effects have been quite small, and many expected radiation effects were not observed. The study to date has revealed no genetic effects (other than the slight and transient shift in sex ratio at birth), no inherited chromosomal abnormalities, no large numbers of congenital anomalies, and only slight growth retardation in these exposed populations. It is possible, as previously mentioned, that further follow-up of descendants of these survivors will reveal significant genetic effects, so that the continuation of the work of the Atomic Bomb Casualty Commission is of great importance for assessing the overall threat of radiation to our species.

Medical Radiation

Since the beginning of this century, an increasing source of radiation exposure to man has come from the use of X rays in medical diagnosis. X rays have proved to be of tremendous value in improving the accuracy and quality of medical diagnosis, so that their use has become increasingly widespread. Few people reading this book will not have had at least one X-ray film.

When X rays are used for diagnosis, the radiation dose to the patient is usually extremely small. However, when the area of these studies includes the reproductive organs (especially if the patient requires repeated X-ray examinations) there is a natural concern for possible genetic effects. As with the study of atomic bomb survivors (whose exposure was much greater), no definite genetic effect has been demonstrated at these dose levels. This is not to say that some genetic effect will not be observed eventually, and it must be stressed again that there is no proof of the existence of a threshold low dose of radiation below which the radiation exposure is safe. Nevertheless, to date, the proven benefits of diagnostic X rays appear to outweigh overwhelmingly their potential harmful effects.

A special case in which this may not be so clear-cut involves diagnostic X-ray studies of the abdomen or pelvis in pregnant women. Here, diagnostic X-ray studies involving multiple exposures to the embryo early in pregnancy have been shown to be associated with a small but statistically significant increase in the incidence of childhood leukemia. As a result, such studies, unless deemed truly critical to the welfare of the patient, are not knowingly performed in women in the early stages of pregnancy or in young women more than ten days following their last menstruation who are not using birth control (and may thus be in the very earliest stage of pregnancy, when the embryo is most sensitive to radiation).

Radiation is also used in another medical setting, as therapy for certain types of cancer. Cancer cells tend to be rapidly dividing, and, as previously mentioned, rapidly dividing cells are generally the most sensitive to damage by radiation. In certain types of cancer, the rate of growth may be decreased by radiation and, in a few types, the patient may even be entirely cured of the cancer by radiation. Therapeutic radiation involves a much larger radiation dose than that used for diagnostic X rays. Consequently, the harmful side effects of radiation are observed much more commonly in this group. Thus an increased incidence of skin cancer, thyroid cancer, and tumors of bone has been observed occurring many years after radiation exposure in areas of the body where patients have received therapeutic radiation. In addition, other types of radiation damage to the skin, intestine, and bone marrow have been observed following radiation of tumors in various areas of the body. Needless to say, in most cases, it can be argued that the potential benefits of therapeutic radiation outweigh the possible harmful side effects, given that the patient already suffers from a usually fatal disease. In cases where there is little hope that the radiation will affect the tumor, and the side effects may prove to be more harmful than the tumor itself, other forms of therapy must be utilized.

Nuclear Power Plants

A relatively new source of radiation in the environment, but one that promises to grow in the future, is the use of nuclear reactors for energy production (Figure 16.5). We will examine the mode of operation of these reactors, how they produce radioactive pollution of the environment, the theoretical exposures to man, and the actual exposures observed to date from operating reactors.

Mode of Operation. The standard type of nuclear power plant uses hollow fuel rods containing uranium as its source of energy. Nuclear fission (a process by which certain atoms are split, giving off large quantities of heat) occurs in these fuel rods. Water circulating between the rods is heated to steam (by the heat of the fission reaction) which is then used to drive a conventional generator, thus producing electrical power. The large amounts of excess heat produced are carried away by a condensor, using a second water supply as its cooling agent. This cooling process, required to

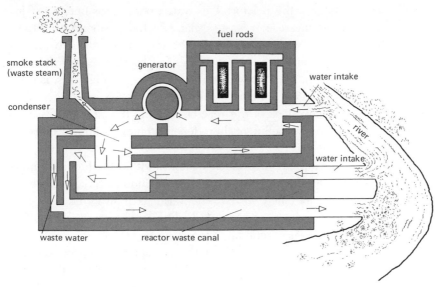

16.5 A nuclear reactor (greatly simplified). Fuel rods in the reactor generate vast amounts of heat requiring vast amounts of water for cooling it. River water brought into the reactor becomes very slightly contaminated as it performs this function. Most of this water, however, is recycled through the system, but a small amount goes into the atmosphere as steam. Another small amount is lost at the condenser. This latter amount leaves the reactor with water which has been brought to cool the condenser and is returned to the river.

prevent dangerous overheating of the reactor, requires very large quantities of water, generally drawn from a nearby river.

Types of Pollution. Radioactive materials are produced as by-products in this system in three ways: (1) The end products of nuclear fission within the fuel rods constitute radioactive waste material which must somehow be disposed of. (2) The protective sheathing of the fuel rods may contain tiny pinhole-sized defects, through which radioactive by-products of nuclear fission may leak into the surrounding water. The system is most susceptible to leakage of a radioactive gas produced in the reactor—krypton-85. (3) The very large amounts of energy emitted from the rods are capable of altering the atomic structure of certain elements in the surrounding water so that they become radioactive. The major product here is tritium, radioactive water.

Possible Sources of Exposure to Man. Theoretically, human beings may be exposed to radiation from nuclear power plants through four general sources: (1) water pollution, (2) air pollution, (3) radioactive waste material, and (4) nuclear reactor accidents.

As noted above, tritium and other radioactive elements are produced in the cooling water of the reactor. This water, once it has been used for cooling, is sent into the reactor waste canal, which deposits it in the river from which the water was initially drawn. Most of these radioactive materials settle to the bottom of the river within a few kilometers downstream.

The water which, upon conversion to steam by the fuel rods, is used for driving the generators contains small amounts of krypton-85 gas. This gas goes up the smokestack of the reactor and into the air. The dissemination of this radioactive pollutant then depends on prevailing air currents in the region of the reactor.

The radioactive products of nuclear fission within the fuel rods must be disposed of when the fuel is spent. These include materials whose radioactivity persists for many years. The waste is transported to a fuel reprocessing center where uranium and plutonium are recovered. The other fission products are generally buried underground in special containers. Along this entire route, from transportation to reprocessing to burial of waste materials, there is the possibility of accidental radiation exposure to man.

The nuclear reactor itself is a potential source of accidental radiation exposure, both to individuals working in the plant and to populations living around it. The reaction occurring in the fuel rods is a controlled form of the same reaction which occurs in a thermonuclear explosion, so that it is theoretically possible, although extremely unlikely, that a reactor could produce a nuclear explosion. Much more within the realm of reasonable possibility is accidental damage to the reactor, with leak of radioactive materials to the surroundings. At the present time, even the possibility of sabotage of a nuclear reactor cannot be overlooked.

Factors Influencing the Amount of Exposure. The potential exposure to man from these sources is influenced by several factors. These include established permissible exposure rates, biological and physical properties of tritium and krypton, and the point in the food chain at which man ingests radioactively contaminated food.

The International Commission on Radiologic Protection has established certain maximum permissible annual doses of radiation to individuals, as indicated earlier in this chapter. For radioactive materials, these have been translated into maximum permissible concentration per year. The concentration of these in the air and water at the boundaries of nuclear reactor sites have been set at a maximum of 10 percent of this permissible concentration. These factors have been monitored by the Atomic Energy Commission.

The biological and physical properties of tritium and krypton-85 influence the radiation dose to man from these materials. Tritium has a rather short lifetime in the human body and its radiation energy is fairly low. Krypton-85 is a biologically unreactive gas and its principal exposure in man is to the skin. Calculations of exposure to man from these two reactor-produced materials have led to the prediction of an extremely low population dose through the year 2000. However, it must always be kept in mind that these calculations are only estimates and may turn out to be inaccurate.

Human radiation exposure from ingestion of radioactive by-products depends very strongly on the length of the food chain between the initial contamination and the ingestion of food by man (Figure 16.6). The longer the food chain, the less the concentration of radioactive materials and the lower the exposure to man. Thus, in a typical food chain, plankton (simple aquatic plants) feed by concentrating minerals in water within themselves. Plankton living in a river used by a nuclear reactor will concentrate radioactive materials. Fish then eat the plankton. The fish, in turn, are eaten by other animals, whose waste material may then fertilize grass or other plants. The amount of radiation exposure to man will depend on where he enters this food chain. The highest exposure might occur to people drinking water from the contaminated river. A somewhat lower concentration of radioactive materials might be ingested by people eating fish from the contaminated river. An even lower concentration would be expected farther along the food chain, for example, in people eating beef from cattle grazing on grass contaminated by animals which fed on fish from the river.

Actual Exposures to Man from Nuclear Reactors. Multiple studies have been carried out in an attempt to assess the actual radiation exposure to human populations from existing nuclear power plants.

Exposures from contaminated drinking water have been studied in the populations surrounding several reactors. A Public Health Service study of the Dresden, Illinois, nuclear power plant showed concentrations of radioactive materials in the reactor drainage canal to be less than 0.1 percent of the permissible levels. These levels would be even further lowered by dilution in the river. Studies of the population of Richland, Washington, downstream from the Hanford reactors (these are of an older design and produce more radioactive waste than modern reactors) showed a calculated maximum annual intake via drinking water of 35 mrem per person (this is well below the maximum permissible dose). In general, it has been found that most of the radioactive materials in rivers used by nuclear reactors are removed by municipal water plants.

Total exposures from fish and drinking water have been measured in several populations. Human exposures from the Clinch River below the

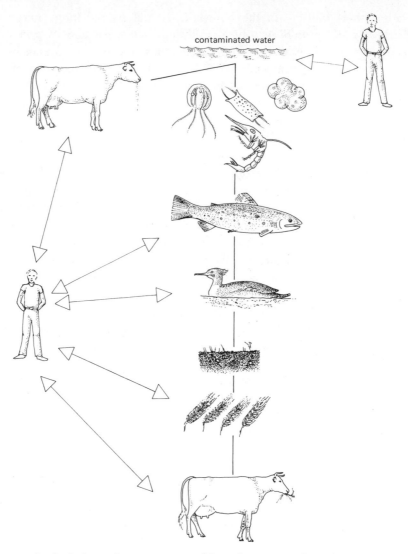

16.6 The food chain when contaminated by radiation. At what point man enters the chain determines the amount of radiation he receives and the potential danger to him. There is a great deal of difference between drinking directly contaminated water and drinking the milk from a cow which might be five steps removed from contamination.

Oak Ridge, Tennessee, reactor and from the Columbia River below the Hanford, Washington, reactor have in both cases been found to be, at most, a few percent of permissible dose levels (Figure 16.7).

Exposure rates to populations from radioactively polluted air are not

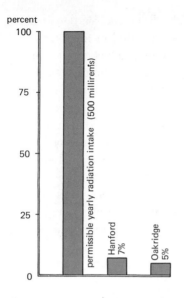

16.7 The maximum one-year radiation intake from drinking water downstream from two nuclear reactors: Hanford (Richland, Washington) and Clinch River (Oak Ridge, Tennessee). The amounts are shown as percentages of a maximum permissible dose of 500 millirems per year as calculated by the International Commission on Radiological Protection. (From data in Sagan, L. A. 1971. *Archives of Environmental Health*. 22:487–492.)

readily measured. Calculations based on an assumption of maximum permissible pollution of air at the reactor site boundary have indicated an extremely low risk from radiation to neighboring populations. These tend to be overestimates, since actual air emissions from nuclear reactors at their boundary sites have almost always measured less than 1 percent of permissible levels. The calculated exposures to neighboring populations have been less than the exposure from naturally occurring background radiation.

A fairly recent study has attempted to look at possible effects of radiation exposure in a population adjacent to a nuclear reactor, rather than at dose levels. An analysis of infant and neonatal mortality rates in areas exposed to effluents from the Shippingport nuclear power plant showed them to be no different after opening of the plant than before the power plant began operating. There was also no difference in mortality rates between areas downstream and upstream from the reactor and between areas with and without exposure to air from the reactor.

In general, studies of the risk of radiation exposure from accidents in nuclear power plants have been somewhat unsatisfactory. No fatal accidents have yet occurred in this country, but the real issue—the risk in the future when greater numbers of nuclear power plants are in operation—is open to speculation. Not surprisingly, two schools of thought have arisen, one claiming that the risks are minimal and acceptable, the other warning that the risks are unacceptably high.

Alternatives: Other Sources of Energy. In the final analysis, the ques-

tion of the potential danger to human health from nuclear power plants
cannot be looked at in isolation. It must be evaluated in the light of the
risk to human health from other modes of energy production and the risk
to human health from dwindling resources of energy. The use of coal as a
source of energy has known detrimental effects on human health, ranging
from very severe lung disease in coal miners to emphysema and an increased risk of lung cancer in populations exposed to air contaminated by
the by-products of burning coal. The dwindling supplies of petroleum and
natural gas as energy sources have not yet produced a major ill effect on
the health of the U.S. population, but this is certainly a future possibility.
The issue is further complicated by the relative financial costs of these
various sources of energy. The cost of nuclear reactors, once thought to be
a rather cheap source of energy, is proving to be more and more prohibi-
tive.

Thus the case for and against the use of nuclear power plants as a
source of energy serves as an excellent example of how complicated problems in human biology and ecology may become. The problem begins
with our incomplete knowledge of the long-term effects of radiation on
human population and the environment and our inability to predict the
magnitude and exact direction of these effects, and is further complicated
by political and economic considerations which have little to do with the
science of biology itself.

The Future:
Use and Dangers
of Radiation

It has been repeatedly stressed in this chapter that our understanding of
the effect of moderate or low doses of radiation on human beings is limited. Thus any predictions concerning the future risks from radiation depend on further development of knowledge in this field. Much of this, of
course, will have to come from animal experimentation, but continuing
follow-up of the descendants of the Hiroshima and Nagasaki survivors may
also yield useful information on the long-term effects of radiation exposure.

With increased understanding of the harmful effects of radiation has
come the possibility of finding ways to protect human beings from existing
radiation in the environment. Several drugs have been developed on an
experimental basis which appear to provide some protection of organisms

from radiation damage, acting at the biochemical level. To date, however, all of these agents have proved to be too toxic for general use by human beings. Still these offer promise for the future, if the toxic side effects can be eliminated.

Finally, the benefits to mankind from the use of nuclear power for energy in the future depend very much on an accurate assessment of the threat of radiation pollution from this source of energy. The already intense debate over the use of nuclear reactors for energy represents one of the major issues in man's attempt to protect his environment—one in which ecological considerations have often come into conflict with economic ones. It should be clear, from the earlier discussion, that there is no simple answer to this problem. As such, it represents an excellent example of the extremely complex problems that face the scientist when he moves from an experimental laboratory setting into the field of human biology and ecology.

SUGGESTED READINGS

Belsky, J. L., and Blot, W. J. May 1975. Adult stature in relation to childhood exposure to the atomic bombs of Hiroshima and Nagasaki. *American Journal of Public Health* 65:489–494.

Gerende, L. J., Taylor, F. H., Sumpter, E. H., Schwarzbach, R. H., and Rogers, K. D. May 1974. Infant and neonatal mortality rates during pre- and post-reactor periods for geographic areas adjacent to Shippingport, Pennsylvania. *Health Phys.* 26:431–438.

Martin, J. A., Jr., and Nelson, C. B. June 1974. Calculations of dose, population dose and health effects due to boiling water nuclear power reactor radionuclide emissions in the United States, during 1971. *Radiation Data Rep.* 15:309–319.

Miller, R. W. October 1969. Delayed radiation effects in atomic bomb survivors. *Science.* 166:569–574.

Pizzarello, D. J., and Witcofshi, R. L. 1972. *Medical radiation biology.* Philadelphia: Lea and Febiger.

Sagan, L. A. April 1971. Human radiation exposures from nuclear power reactors. *Archives of Environmental Health.* 22:487–492.

Shapley, D. June 1971. Radioactive cargoes: record good but the problems will multiply. *Science.* 172:1318–1322.

17
THE ECOLOGY OF
SIMPLE SOCIETIES

The destiny of a human group rests on its dynamic interaction with the environment; its ecology. In present popular usage *ecology* tends to be confused with the environment itself; actually it refers to the continuous exchange, between the environment and the group, of material and energy which the group needs for its existence. The environment involves climate and geography as physical agents; plants, animals, and men as biological agents; and by no means least, man's distinctive product, culture.

Various adjustments or adaptations are necessary for successful existence in a particular environment. Each member of a group has the ability to adapt, to a greater or lesser degree and so has the group as a whole. The natural environment may be partly or largely replaced by a domesticated environment of shelters, gardens, and farmlands; a higher stage is industrialization and urbanization.

Demes and Ecosystems

Human groups vary greatly in size and organization. The simplest, relatively isolated, self-sufficient group would correspond to the deme, the natural local population among animals or human beings (Figure 17.1). Larger groupings of man range from tribes, which are complexes of demes, to tribal associations, which may be distinguished by a common language. Then come nations and groups of nations.

An ecosystem is a complete functional natural system, of any size. It

17.1   Idealized human demes: semi-isolated self-sufficient villages.

contains all forms of animal and plant and their interactions with one another. The ecosystem also contains the physical environment, which supplies the nutritional needs of the animals and plants and is affected by the living organisms as they exploit it. (Soil formation, for example, depends on a multitude of small animals and microorganisms plus vegetable refuse.) Within an ecosystem there operate the evolutionary forces which change living species or diversify them.

An ecosystem considered from the human viewpoint sees man and his needs as part of such a system, adapting to it or affecting it. (This is, necessarily, the viewpoint used here, not that of ecology in general.) The human species is basically a widespread network of demes, or natural populations, and overlapping ecosystems. Demes are temporary, in that the individuals composing a deme in a particular ecosystem can merge with other demes or can split off to found new demes.

Ecological Adaptations

Biological Responses. Environmental stimuli are of two kinds: those shared by all men, as a species characteristic, and those distinctive of a few individuals or populations. All normal people respond to altitude by increasing the rate of breathing, and respond to heat by sweating. This physiological plasticity of individuals is an important characteristic of the species, developed in the course of evolution. It ranks in importance with the generalized structure of the human limb or the ability of the human brain to think in abstract, symbolic terms.

The second kind of biological adaptation comprises features shared by small numbers or by distinctive human populations. Examples are the long, narrow nose of people living in dry areas, or the distinctive type of insulative response to cold found in Bushmen or Australian aborigines, in which they permit their extremities to become colder than the central body.

Cultural Responses. These consist of those social and technical activities which permit the group to come to terms with or to dominate the environment. They include social institutions such as organization, division of labor, marriage and child-rearing customs, as well as technical provisions for obtaining food, shelter, and clothing.

Human Ecosystems

Habitat Types. Human ecosystems can be classified in terms of habitat and type or complexity of economy. Habitats are roughly classified into ten groups, nine of which depend on geographic latitude; the tenth, mountain habitats, depends on altitude (Figure 17.2). The first nine are as follows: deserts or drylands; tropical forests; tropical scrub; tropical grasslands (savannah); temperate forest; Mediterranean scrub; temperate grasslands; boreal land; and polar lands and tundra. Each has characteristic vegetation and animal life and each can be found on all continents. Man originated in tropical grasslands or scrub lands. Modern man has been more successful in living elsewhere, as shown by the high population densities in temperate climates and even in tropical forests.

Economy Stages. The type of economy can be classified according to the major productive technique used. The earliest and least productive economy is hunting and gathering. Fishing is of course one type of hunting. Herdsmen and nomads practice a special kind of economy, based on cattle, sheep, or goats, Simple or shifting cultivators practice simple garden agriculture, supplemented by gathering, hunting, and domestication of pigs and poultry. Advanced agriculturists rely on intensive grain agri-

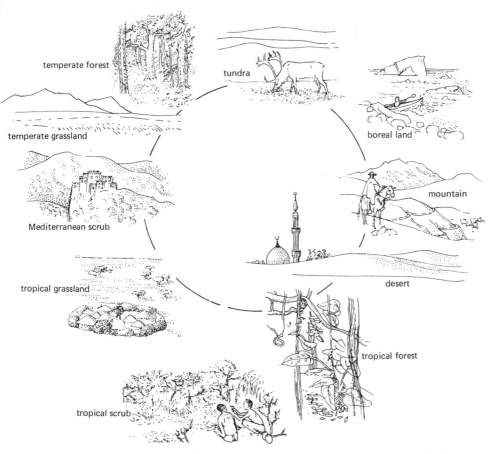

17.2 Natural habitats. Note that culture often provides man with a variety of adaptations.

culture as their main means of subsistence. This development, only nine thousand years old, was probably the most important single step in the cultural evolution of modern man (see Figures 15.3 and 15.4). It made possible large populations, cities, civilization, specialization, and all the biological and social consequences that we have been discussing in this book.

For a given level of technology, some habitats are obviously more favorable than others. In fact, some environments will not permit certain economics. For example, a cattle herdsman or agriculturist would not get very far in the arctic. But most habitats permit a variety of economic stages and settled agriculture permits high population density in most human habitats.

Measurement of
Ecological Adaptation

How can we measure the ecological adaptation, or success of a group? We can apply demographic or medical standards, such as population size and its maintenance, rates of death and disease, life expectancy, longevity, and nutritional status. Depopulation, of course, is the most definitive index of maladjustment of a population to its environment. A migrant group may fail to establish itself. Or, in many Pacific islands, depopulation occurred as a result of sterility due to gonorrhea introduced from the outside. Sometimes a particular environment may be marginal or precarious in sustaining human life, as in parts of New Guinea, with poor soil, deficient minerals, and very little wildlife. In the New Guinea highlands, however, skillful gardening and a healthy climate have led to a relatively heavy population for primitive villages. Tropical Africa has always been a difficult environment, combining climatic stress, nutritional deficiencies, and many infectious diseases.

If a society or human group can perpetuate itself and function, this very fact demonstrates genetic and cultural adaptation. Energy input and expenditure must be in reasonable long-term balance; otherwise, the people would be starving and dying. Of course, too much intake and too little expenditure, as in affluent societies like our own, can also be harmful, resulting in obesity as a warning signal and then in increased mortality from diabetes, high blood pressure, and coronary heart disease. But this situation for entire populations is very new in human history. Even now, it affects only a small fraction of the world's people. Most of the world would be delighted to exchange their undernutrition for our overnutrition.

Anthropometry. Many groups function remarkably well with diets and with hemoglobin levels that would be considered grossly inadequate in the United States. But the level or quality of their adaptation will vary. Two simple techniques, anthropometry and hemoglobinometry, can indicate the level of adaptation. Hemoglobinometry is a direct measurement of red blood cell mass, which reflects the state of nutrition, particularly iron intake. Anthropometry means body measurements, like height, weight, and skinfolds. Whether applied to child growth or to the comparison of adults of different ages, in order to detect a long-term or secular trend, these simple body measurements can detect degrees of malnutrition far more subtle than any known clinical examination or biochemical test—including the measurement of hemoglobin! By comparing the growth rates of an individual child, or of a group of children, to the growth of well-nourished children of similar genetic backgrounds, we can show the need for health programs, such as dietary improvement, malaria eradication, immunization, sanitation, and the like and we can measure the amount of improvement. The date of any improvement in health can be estimated retrospectively (that is, long after the fact) by finding a secular trend toward increased stature or earlier menarche. Any retrogression in health, such as during wars or famine, can be similarly detected and dated. Secular increase in height shows that the health of African Bushmen has been improving over the past seventy years. The same is true for various other groups, even those with little contact with the outside world, beyond the adoption of steel tools. Steel tools make hunting, gathering, and farming more efficient, and very few people in the world remain without such tools.

Behavior. One of the great needs in assessing the quality of a group's adaption is for an objective measure of behavioral stress. In other words, are the people happy? Do they show a high or low level of disturbances of mood, thought, or feelings? How does their rate of neurotic symptoms or their amount of antisocial behavior compare with that of other groups? Apparently, the frequency of major mental disorder, or psychosis, is fairly constant in all populations and at all times, at about 1 percent (Figure 17.3). This is the rate in the U.S. Navy in 1860 and 1960, in aboriginal villages in Taiwan, in rural Nova Scotia, in urban Nigeria, and in midtown Manhattan. Such constancy, we might note parenthetically, suggests a biological component to major mental illness. Twin studies point in the same direction. But to return to the milder forms of behavioral disturbance, these have been studied cross-culturally by Alexander Leighton, at the Harvard School of Public Health, using a standard interview technique and trained native observers. On the whole, he found similar patterns of distribution of psychiatric disorders in rural Nova Scotia, rural Nigeria, and urban Nigeria, with a somewhat lower overall frequency in Nigeria. Women showed more psychiatric disorder than men in Nova Scotia, but less in Nigeria. Leighton attributes the sex difference to the differences in the impact on the two sexes of social determinants of stress in the two cultures. Finally, Leighton found that sociocultural disintegration had a harmful effect on mental health.

This kind of study should be greatly extended to many more groups, particularly those which are least touched by civilization. It is important to know whether primitive man is noble and happy, as Rousseau thought, or whether Hobbes was correct in terming the lot of early man "solitary, nasty, brutish, and short." Some groups in New Guinea are under con-

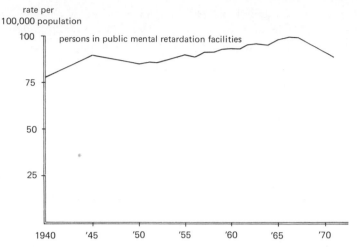

17.3   Mental retardation rates: 1940–1971, residents of public mental retardation facilities. (From *Social Indicators*, a publication of the Social and Economic Statistics Administration, U.S. Department of Commerce, 1973.)

stant stress, fearful of ambush or sudden death in blood feuds, perennially quarreling, violently hostile to outsiders, and barely civil to their wives. And yet these people have low average blood pressure. Blood pressure is a crude index of stress, applicable mainly within developed societies. As I have pointed out, primitive people generally have a different pattern of disease from advanced or industrialized people—more infections and fewer diseases of civilization. I will return to this.

Nutrition. Evidence has been accumulating that hunters and gatherers are fairly well off nutritionally. Even in desert environments, dietary resources are diverse. Richard Lee has shown that the Bushmen spend, on the average, only two to three days a week getting food and do quite well. Colin Turnbull's account of the African forest pygmies shows that they enjoy an adequate diet, although they work harder than the Bushmen. On the other hand, the Siriono of the Bolivian rain forest are always at the edge of starvation. The anthropologist who studied them found, in fact, that unlike most human groups their dreams revolved around food. Freud pointed out that dreams can represent wish fulfillment and most people in Europe and America dream not of food but of other well-known appetites.

Another exception to the generalization that hunters and gatherers are fairly well off nutritionally have been the Eskimos, whose adaptation used to be the most precarious existence on earth. Whereas most hunters and gatherers are subject to rare catastrophic failure of food supply due to drought or other climatic factors, Central Arctic Eskimo in the past courted this disaster continually. There was very little vegetable food in their diet for long periods of the year. The kinds of animals that are dietary luxuries to many hunters and gatherers were, to the Eskimos, the absolute necessities of life. At present, of course, those Eskimos who are not living on governmental distribution of packaged food have become spectacularly successful hunters because of modern technology in the form of power boats, explosive harpoons, firearms, and snowmobiles. In fact, they are beginning to deplete their supply of fish and game, which poses a whole new set of problems for the future. But for the present they are doing very well.

On the whole, it is the agriculturists who have the most assured source of food supply. However, when farming only one crop, they risk catastrophe in a crop failure, whether due to climate or a disease affecting their single crop. The Irish potato famine is a classic example. Another example was the taro blight in the 1950s which virtually eliminated that food source in the western Pacific. Fortunately the sweet potato was available as an alternative and has now become the staple foodstuff in Melanesia. One can imagine the tragedy if some of the newer strains of rice in the Orient (Figure 17.4) should suddenly fail, as certain strains of hybrid corn did during the summer of 1970 in the United States. In that year, 10 percent of the entire corn crop was lost, due to a new mutant form of a virus.

17.4   Land is not a resource to be wasted in highly populated Indonesia; terraced rice paddies, Bali. (courtesy Peter Timms)

The loss reached 50 percent in some states. This is the other side of the Green Revolution—the risk that if the new grains replace the old and if something should happen to the new ones, starvation could become widespread within months. It shows the need for preserving a variety of strains for each crop, in a genetic bank, to protect against such catastrophes.

The nutritional state of primitive groups depends on the fertility of their habitat, being good among the African pygmies and Bushmen, or Oceanic fishermen, for example, and poor among the Sandawe of Tanzania or the Siriono of Bolivia. But culture also determines the extent to which the environment is exploited. In Chapter 1, I cited the case of two Solomon Island communities, both living on small islands or atolls in waters abounding in fish. One, on Ulawa, is good at raising sweet potatoes but culturally indifferent to fishing and poor at it, and an appreciable amount of malnutrition exists. The other, on Ontong Java, is made up of avid and expert fishermen who thus add plentiful protein to their taro and coconut diet; and there is virtually no evidence of malnutrition. Another example of undernourishment because of a cultural aversion to making use of plentifully available fish is reported for tribes living along the Red Sea.

Outside of nutritional deficit in poor natural areas, most undernutrition occurs in settled agricultural or peasant societies, as in Asia and Latin America. Here, man's very success in raising food has permitted population increase, which has outrun the increase in food supply.

The Biological Price of Progress

The shape of the population pyramid in primitive cultures is generally intermediate between the broad-based pyramid (many children) of settled agriculturists in underdeveloped regions like Latin America or India, with high birth and death rates, and the tall, narrow pyramid (larger proportions of mature and old) of advanced societies, with low birth and death rates. Primitive populations in the natural state seem to have moderate birth and death rates. Except for Oceanic atolls like Tikopia, the numbers of simple peoples are usually not limited by the amount of available land but by its carrying capacity. Thomas, in fact, proposes that population size is determined by balancing the universal desire for children against the extra work needed to feed them in a given environment. Once nonindustrial man survives the childhood infections and reaches adult life, his

remaining life expectancy is closer to that of technologically advanced man
than is true of his life expectancy at birth.

Demography. Primitive populations are demographically young.
Their ratios of persons under fifteen to those over forty-five are less than
the very high ratio of agricultural peasants, as in India or Latin America.
Very high ratios are also found among recent recipients of massive public
health intervention, such as Ceylon (malaria control). The lowest ratios
are found in the advanced countries, such as the United States, Western
Europe, and Japan. Specifically, the ratio of persons under fifteen to those
over forty-five in the United States is about 1.0, as it is for other developed countries. Pacific islanders vary from 2 to 3, while the ratios in india
and the island of Mauritius, in the Indian Ocean of East African run from 3
to 4. The preponderance of young people means, of course, that the population is increasing rapidly. People with the simplest economies generally have moderate fertility and mortality, whether they are hunter-
gatherers or shifting agriculturists. Fertility is regulated culturally, by
taboo on marriage and on intercourse during lactation and at other times.
Abortion and infanticide, while occasionally practiced, are on the whole
uncommon among tribal folk. With the introduction of a major world
religion, such as Christianity or Islam, these customary taboos break
down, increasing fertility. The dominant culture abolishes the former
social determinants of mortality, such as abortion, infanticide, cannibalism, headhunting, killing of old people, blood feuds, and other
kinds of warfare. [Figures 17.5 to 17.7]

17.5 Sitting Bull addressing government officials about land claims (July 1888).
(courtesy Peabody Museum, Harvard University)

17.6 Culture clash: two subtle weapons on the white man's side—religion and education. Indians from the Lower Fraser River being led in prayer. (courtesy Peabody Museum, Harvard University)

17.7 Indian students at the Carlisle School, Carlisle Pennsylvania. (courtesy Peabody Museum, Harvard University)

Exposure to the larger culture, then, weakens traditional taboos, whether by direct missionary influence or by observation and imitation. When, in addition, modern medicine (Figure 17.8), public health, and improved nutrition become available, the combination causes an explosive increase in population. Such has been the experience, for example, of the

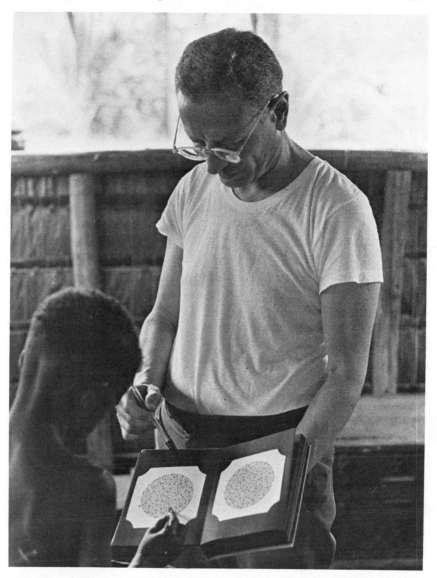

17.8 Dr. Albert Damon testing young boy for color blindness. Solomon Islands, 1970.

Tolai in New Britain, a group whose numbers have increased so fast since World War II as to require massive resettlement by the government on formerly empty lands. They have outrun their means of subsistence. This is a very real, although as yet insufficiently appreciated danger for many primitive peoples. Most attention to overpopulation has so far been concentrated on the single-crop, settled agriculturists of undeveloped countries like India or Latin America. But the Samoans, to take another example from the Pacific, have increased some tenfold during this century, with a recent doubling time of eighteen years.

Medical Effects. Westernization shifts mortality and morbidity away from the indigenous infectious and parasitic diseases, sometimes complicated by malnutrition. Tribal man carries a considerable load of parasites, inherited from our primate ancestors and evolving in step with man. These parasites are both internal and external. The first shift is, of course, the arrival of new infections brought by the civilizers. The classic examples are the introduction of veneral diseases, smallpox, and tuberculosis by Europeans to the New World and Oceania, and of leprosy by Chinese throughout Australia. In fact, current mortality and morbidity statistics from the Solomon Islands now resemble those of the United States in 1900. It is intriguing that accidents rank as high as the third leading cause of death in both the Solomons and the United States today. One might expect that technology would increase the accident toll, relative to other conditions, but apparently it does not.

With increasing acculturation come the beginnings of the chronic, noninfectious conditions that predominate in the developed countries. Burkitt cites several disease complexes characteristic of technologically advanced societies but rare in simpler ones, notably diabetes, obesity, coronary heart disease, and high blood pressure; and his own postulated "lower bowel complex" which includes appendicitis, ulcerative colitis, diverticular disease, polyps, and cancer. The civilized diet, soft in texture and high in refined sugars, produces dental cavities. Disharmonies of the jaws, or malocclusion, requiring orthodontic treatment in developed countries, join dental cavities to constitute a dental complex as another disease of civilization. And finally, our work in the Solomon Islands may have established (by its low incidence there) an ocular .complex—astigmatism, myopia (near-sightedness), glaucoma, and color blindness.

Genetic Effects. We have already noted some of the ongoing evolutionary trends in modern man, the current into which preliterate societies are being swept. Such trends include the breakdown of breeding isolates; selection for adaptation to new, man-made environments and to new diseases; relaxed selection against the traits no longer harmful; environmental amelioration, with consequent full expression of genotypes and an increasing load of genetic defects; and the coexistence of stabilizing with directional selection. Let us now consider a few immediate genetic mechanisms and consequences of the transition from primitive to cosmopolitan ways of life.

James Neel has drawn attention to the drastic change in selection intensity when fitness, or reproductive success, no longer corresponds to ability. In many primitive groups outstanding men, whether hunters, planners, or artists, have more wives and children than other men. Among the Yanomamö, a South American Indian tribe, fewer than 4 percent of the men contributed 23 percent of the tribe's grandchildren. Selection intensity, being proportional to the variation of the number of children per person, is very high in the Yanomamö. Contrast this with the situation when polygyny is forbidden, when customary limitations on reproduction are discarded and when most children survive—all of which can occur within one generation after culture contact.

Another intriguing question is the variability of inbred isolates in polygenic traits. Paradoxically, inbreeding should increase variability, by increasing homozygotes (which are apt to be further from the average) at the expense of heterozygotes. Neel has, however, found no such increase in metric and morphologic traits among Brazilian Indians, nor have R. C. Romba and I in unpublished research on height and weight in several primitive and cosmopolitan populations. Range restriction cannot explain our failure to find the expected increase, nor should fixation by random gene loss work in this way for polygenic traits.

One possible mechanism for the seeming constancy of metric variation is stabilizing selection. As already mentioned, we have found evidence for stabilizing selection in two relatively unacculturated Solomon Island tribes, where men of midrange height had more children than shorter or taller men. (Assortative mating would be expected to increase variability in the same way that inbreeding does, but such mating by stature has not been found outside of Western European and American populations. However, some Solomon Island groups do, as we saw, mate assortatively for skin color.)

SUGGESTED READINGS

Coon, Carleton, S., ed. 1948. *A reader in general anthropology.* New York: Holt.
Coon, Carleton S. *The hunting peoples,* 1971. Boston: Atlantic Monthly Press.
Forde, C. Daryll. 1934. *Habitat, economy and society.* London: Methuen and Company.
Lee, R. B., and Devore, I. 1968. *Man the hunter.* Chicago: Aldine.

18
THE BIOLOGY
OF URBAN MAN

Hunting and fishing for food; herding sheep and cattle under the open sky; growing food in green valleys or coral gardens; this may have been the Golden Age of man. It is a nostalgia which grips many who live in the complexity of a modern nation and even takes a few of them out and back into such simple existences. But the rest of us are doomed to the new ecology, that of the city, invented about three hundred years ago; that is the industrial city, not the temple centers of Mesopotamia or the Maya, nor Athens of the Classical age, nor the London of Elizabeth I. This kind of city is now growing up in every country, stretching, for example, in the United States from Portland, Maine, to Norfolk, Virginia, without a real break (Figure 18.1), or all around the Great Lakes in rapidly spreading patches. This is the last great revolution in man's history: the cities are inexorably sucking the human population into them like whirlpools.

Pretechnological Living

But we may lay to rest the myth that primitive man had a relation to nature different from our own, and that exploitation, or perhaps ravaging, of the environment is something new. Man has always modified and disrupted the world around him, restrained only by the level of his technology. In the last two generations, earth-moving machinery has replaced the puny pick and shovel, which nevertheless built railroads across the United States.

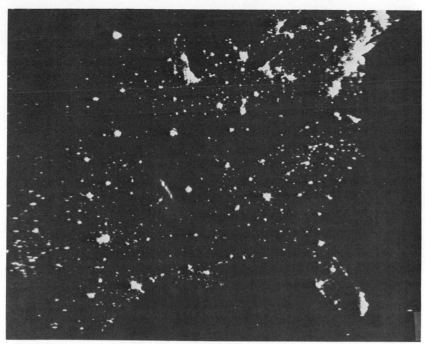

18.1 Dense urbanization of the eastern half of the United States is recorded in this infrared photograph taken at night by an Air Force weather satellite. Patterns of heat concentration reflect the distribution of population. (courtesy USAF weather satellite)

Primitive and modern societies, for purposes of this discussion at least, differ in three significant ways: population size, level of technological sophistication, and attitude toward the natural environment.

The populations of primitive people were small. Their waste products were for the most part biodegradable and did not occur in such concentrations as to overwhelm the ability of natural systems to deal with them (Figure 18.2). The situation is far different today. Even our natural wastes are so large in amount that they are accommodated by natural processes only with difficulty. In addition, we turn the resources of our environment into new synthetic or plastic substances that cannot be degraded by natural processes. Furthermore, our technological sophistication allows a much greater level of consumption than was possible for primitive man. This very real difference in ability to affect the environment must not be confused with a difference in attitude toward the environment.

Primitive Societies as Consumers. Despite the kinship with nature that primitive man may have felt, his actions were another matter. The Sioux Indian who would not drive stakes in his mother, the earth, or cut her with a plow, showed no qualms about driving a herd of buffalo over a

18.2 In prehistoric and simple socities, little is accumulated which will not return to the earth. Lone Wolf's Camp. (courtesy Peabody Museum, Harvard University)

cliff, or about starting a range fire to stampede them. The Indians, like the wolves, often ate only the choicest part of the buffalo, the tongues, when game was plentiful, and left much of the kill unused. In fact, the hunting abilities of primitive man are believed to have caused widespread extinction of large mammals during the Pleistocene Epoch, which includes the last three million years (Figure 18.3). This has been called "Pleistocene overkill." The introduction of the horse made hunting so efficient that the Plains Indians would soon have exterminated the buffalo by overhunting if the white man had not overhunted them with firearms. The cliff dwellers of Mesa Verde, in the Southwest, threw all their garbage over the edge of the cliff, just as men did in medieval cities. Modern man behaves somewhat better, except on picnics or at football games. The Navaho, in adopting sheep, contributed to the overgrazing of the Southwest.

The point is that whatever the Indians' attitude toward nature, their actions were and are identical to those of modern man. Throughout his long history, man's only concern for the environment has been what it could contribute to human survival.

The Rise of Urbanization. The movement of populations into always

18.3 Some Pleistocene extinctions. Cave paintings of animals hunted during the Palaeolithic of southwestern France and which are now extinct. Cave bear, mammoth, horse, woolly rhinoceros, reindeer, lion, and wild bison.

larger units has been a feature of the human scene ever since settled agriculture made it possible to feed large groups of people in one place (Figures 18.4 to 18.6). The process of urbanization has accelerated with time. All over the world, people are leaving rural areas for what they see as the

18.4   Palaeolithic rock shelter, southwestern France. (courtesy Peter Timms)

18.5   Roman City, north Africa. (courtesy Peter Timms)

18.6   Village, southwestern France. (courtesy Peter Timms)

excitement, the opportunities, and generally the better life in cities. And on the whole, even after some disappointments and disillusionment, they stay, and do not return to the farm or village (Figure 18.7). Anyone the right age remembers the comic World War I song: "How ya gonna keep 'em down on the farm, after they've seen Paree?" This has been getting less comic ever since.

Some form of housing is very old in human history (Figure 18.4). The hominids were apparently already protecting themselves from the weather by the use of windbreak shelters and caves one million years ago. A well-developed skin shelter has been traced out in a cave on the French Riviera, dating to the height of the next-to-last glacial phase, perhaps 350,000 years ago. Men began to build permanent dwellings, recognizable as houses, as they settled down to agriculture before 7000 B.C. Houses and villages appeared at essentially the same time. The first cities appeared five to six thousand years ago.

From the beginning, the city provided so many advantages to men that it became the center of cultural evolution. When the earliest cities appeared, probably not more than a fraction of 1 percent of the world's

18.7 The slums of Bombay. In many countries, people from rural areas are flocking to the cities faster than the city can provide shelter, services, or jobs for them. (courtesy Warder Collection)

population lived in them. Today, the figure is 19 percent, but that is for the world as a whole (Figure 18.8). The proportion of urban dwellers in each nation is roughly parallel to the level of technology and the standard of living of that nation. In the United States and Canada, 70 percent of the

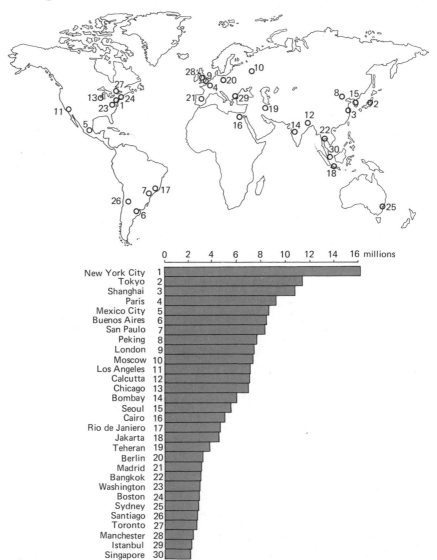

18.8 The population of some of the world's largest urban areas. (From *The World Almanac and Book of Facts*. 1976 edition; Copyright © Newspaper Enterprise Association, New York, 1975.) Some of the figures appear to be conservative. Mexico City, for example, now has approximately ten million inhabitants, while its greater metropolitan area is said to total fourteen million.

people live in cities, in Brazil, 46 percent and in India, 15 percent. The percentage is increasing steadily everywhere. Very recently, in the past twenty years in this country, the suburbs have been growing faster than the cities, but this does not affect the main argument.

Man and the Environment

Human life, like every other form of life, is based on a constant interaction with the environment. There are environments in which men thrive, others in which they sustain themselves with difficulty, and still others in which they cannot live at all. One might think, then, that any effort made to preserve the quality of the natural environment, to enhance its beauty, and to protect the other living things that are found in it, would lead to an improvement in the quality of human life also. Unfortunately, this is not the case. Over the long run, men are utterly dependent on the integrity of the natural environment; but over the short run, there may be a direct conflict between the need of men and the desires of those who wish to maintain the integrity of some parts of the environment.

The Biological Interfaces. Man's environment begins at the boundaries of the man—at his skin, at the lining of his lung, and at the lining of his intestinal tract. These, and his sensory organs—eyes, ears, olfactory cells, taste buds, and the specialized nerve cells in the skin, which inform him of touch, pressure, temperature, and pain—are the major interfaces. The clothing on his back, the air in his lungs, and the food in his stomach are not a part of the man, but a part of the environment. The people around him are also a part, and a most important one.

The statement that life is based on the constant interaction with the environment is literally true for every one of us. Cut off a man's air for a few minutes and he dies; expose him unclothed to a cold winter's day for a few hours and he dies; cut off his water for a few days and he dies; cut off his food for a few weeks and he dies; cut off his patterned sensory input, by floating him blindfolded and ear-stopped in water at body temperature, and he begins to go to pieces mentally within three days; put him in a prison cell and isolate him from all contact with his fellow men and he goes to pieces mentally within six weeks or so.

There is a widespread belief today that modern men live in an environment not natural for them, and that they are partly insulated from their environment by society. Nothing could be more untrue. Men who live in modern societies are as totally and critically dependent upon their environment as any other living organisms. The question is not whether men are insulated from their natural environments; rather, the question is, "what is the natural environment of modern men?"

The Modern Natural Environment. Much of this line of argument, which is certainly unorthodox, has been developed by Dr. L. E. Hinkle of Cornell Medical School. His answer to the question above would be as follows: In their natural environment, modern men live in highly complex social groups, organized as national or supranational states. They wear clothing. They live in man-made dwelling units, within which they maintain a warm, dry, microclimate. Most of them inhabit cities and towns. Most of their food is derived from agriculture and animal husbandry. Some of it is manufactured. A great deal of it is extensively processed and cooked before they eat it. Modern men spend most of their time interacting with other members of their own species, and utilizing complex machines. They have remarkably rapid means of communication and transportation which are an integral part of their societies. They are aggressive and potentially dangerous animals, but they are very intelligent and very adaptable.

Now Hinkle admits that the "modern men" he is describing are creatures of the last three hundred years, and that their "natural environment," as he describes it, is artificial in one sense. It is the outgrowth of modern technology. But, you may ask, is not the true natural environment of man the forests of Europe and Africa, through which our ancestors roamed naked a few generations ago? Is not this a natural environment in which men might thrive, free from all the ills associated with modern society?

To this Hinkle would reply, Modern men thrive in modern environments such as the one just described. In more primitive settings, reminiscent of earlier stages of their development, they do less well. Thrown naked into a forest, without tools, with no prepared foods, and with no protective dwellings, they would not survive. Of course, a few weeks' training in Outward Bound or the Boy Scouts might prevent this.

Let us consider some of the biological characteristics of our species. Modern men are the products of culture and of the social group, and the development of both preceded the appearance of our present species. The social group is part of our biological heritage which we share with many other vertebrates. The cultural adaptation was evolved by earlier species of Hominidae long before our own species appeared.

For over two million years before *Homo sapiens* evolved, other hominids had been using tools and living in hunting bands. At some point language appeared. For at least five hundred thousand years men had been using fire, cooking their own food, and living together in small bands. Clothing and the building of shelters came in the Ice Age, perhaps before modern man existed, and these ways of living were important factors in the evolution of our own species. It was not the other way around; we, that is, modern man, did not invent them. However much man may have shaped his culture, it has also shaped him to an importar.t degree, in a truly biological sense.

Man in the Food Chain. The precultural hominids were hunters and scavengers, who obtained all of their energy from food. In technical terms, they were members of a terrestrial food chain in which they were tertiary consumers—carnivores who ate other carnivores, as well as herbivores and green plants. They, in turn, were prey to other macroconsumers—large carnivores, including other hominids. They were also prey to microconsumers—bacteria, fungi, plants, and small animals, many of which were parasitic upon them. As they acquired the ability to use tools and to make weapons, the hominids effectively fought off the large carnivores that occasionally preyed upon them. Although cannibalism persisted in some communities until historic times, and even now occurs sporadically, modern men have not been a major article of food, even for members of their own species, for many thousands of years. By contrast, it was not until the development of modern technology that men even recognized the existence of the micropredators, which have always been their chief biological threat. Once the role of microorganisms in human disease was understood, vigorous and effective methods were undertaken to control them, methods that have been enormously successful. During the last hundred years we have made considerable headway in eliminating ourselves as a source of food for microorganisms, at least while we are alive.

As soon as men developed efficient weapons, and more sophisticated techniques of hunting and fishing, they began to make their own food supply more reliable. Then, we saw, their major advance in this area occurred with agriculture, as they started to domesticate other animals some ten to fifteen thousand years ago. They have made a still further advance in recent years with the development of artificial fertilizers, genetically scientific plant and animal breeding, weed and pest control, and the mechanization of agricultural labor. Agricultural productivity has now reached the point at which a single farmer in the United States can feed himself and forty-five other people.

The Control of Nonhuman Energy. So men have markedly altered their relation to biological energy from food. When *Homo erectus* began to use fire, some five hundred thousand years ago, the hominids began to obtain energy from sources other than their food. At first the amount was very small—enough for some heating, cooking, and driving game animals during hunts. When *Homo sapiens* began to domesticate other animals, he obtained additional sources of energy for pulling, hauling, and riding. By the time the first city-states were developed each man in these societies was probably putting into play from other sources as much energy as he obtained from his own food—that is, about three thousand calories a day.

These extra-metabolic sources of human energy were gradually augmented over the years by the use of wind and water power; but there was no considerable addition to them until the development of heat engines powered by coal, at the beginning of the eighteenth century. Since then they have grown exponentially. In 1860, when the population of the United States was thirty million, each man was receiving from extra-metabolic sources an amount of energy equal to that of six other men. A century later, in 1965, when the population had increased sixfold, the energy supply per person had increased twelvefold. By then, every individual in the United States commanded an energy supply equal to that of seventy men. All of it came from sources other than food—95 percent from fossil fuels, coal, and oil. We could not maintain our present civilized lives without it.

Civilization
and Health

A rich and extensive culture and complex social groups are natural features of modern man. He is never found without them even in primitive cultures. What has been the effect of all of this on human health? A highly beneficial one, both from the development of civilization during the past ten thousand years, and from the technological explosion of the last three hundred years.

Much is known about the health of prehistoric hominids, learned directly from the study of skeletons and mummified bodies. We have also learned it indirectly from the study of surviving primitive societies, as we saw. By modern standards, the health of prehistoric men, like the health of men in primitive societies, was very poor indeed. The average life expectancy of prehistoric men has been estimated to be about twenty years. Infant and childhood mortality was very high. Only a minority of the children lived to the age of ten, most of them dying, in great numbers, of birth injuries, infections, malnutrition, and trauma. Those who attained adulthood often died early of injuries and infection. Rarely did anyone live to the age of fifty.

The reasons are clear. Living intimately as a part of the natural system, prehistoric men and men in the most primitive of recent societies were subject to all of the assaults of that system. Being part of a natural food chain is no fun for the participants. From the moment of birth, prehistoric men were dependent on a limited, capricious, and grossly contaminated food supply. They were infected by bacteria and viruses, infested internally with worms and protozoa, and externally with lice, ticks, mites, and other arthropods. Infants born to chronically infected and marginally nourished mothers were subject to congenital defects, birth injuries, exposure, and innumerable infections.

Despite this, some of the early human groups of hunter-gatherers and some of the primitive agricultural societies made remarkably good adaptations to their surroundings. In relatively isolated and unchanging settings, they might develop an equilibrium with other parts of the natural system. As we saw in the last chapter, this equilibrium is based on cultural controls of the rate of reproduction, a relatively (although not absolutely) low infant and childhood mortality rate, and a reasonably adequate food supply. In such a group, those who survive to grow up may attain a remarkable degree of symbiosis with their parasites which allows them to be vigorous and hardy, and to live to an old age. Adults beyond the age of fifty might number as many as 10–15 percent of such a group.

These primitive groups nevertheless remain highly vulnerable to any change in their natural environment which might threaten their food supply. Historically, they have become ill and died in large numbers when contacts with people from the outside exposed them to new microorganisms to which they were not adapted. Also, when members of these groups have been subjected to new patterns of living, their adaptations to their own microorganisms have broken down, and they have suffered the catastrophic consequences in illness and death.

Agriculture and Its Effects. In a way, the development of agriculture was the most important public health measure in the history of mankind. Agriculture markedly disturbed the balance of nature wherever it was practiced, but it dramatically improved the human food supply. In ten thousand years, from 8000 B.C. to 1650 A.D., human life expectancy rose irregularly from about eighteen years to about thirty-five years, and the population of *Homo sapiens* increased from perhaps five million to about five hundred million. As societies became more settled, infections continued unabated. New diseases became prominent, particularly the great epidemics, such as the plague, typhus, cholera, and smallpox, epidemics made possible by the aggregation of large numbers of people living in close association with each other and with their domestic animals, as well as with rats, fleas, lice, and flies. Under such conditions everyone shared bacteria, viruses, protozoa, and worms with everyone else. Over this period of pretechnological civilization, the health of men changed very little. But since 1650 it has improved markedly, which can be attributed directly to the explosive rate of technological development that has taken place and to all of the changes in living conditions that have accompanied it.

Impact of Industrial Technology. There have been two notable factors in this improvement in health. One has been an improvement in the human food supply, made possible by the increase in the productivity of farms, the improvement in the methods of processing, preparing and preserving foods, the improvement in transportation, and the development of the nutritional sciences. The second has been the discovery of the role of microorganisms in disease. This has made it possible to protect men from infection by sanitizing water and food supplies, by disposing of human feces, by attacking insects and other living vectors that transmit disease, by attacking the animal hosts that are its reservoirs, and by developing chemotherapeutic agents and antibiotics which attack the microorganisms, but not the person who is infected. Note carefully: All this medical activity has disturbed the balance of nature, but it has been highly beneficial to mankind.

In 1650 human life expectancy in Great Britain at birth was about thirty-two years; in 1850 it was about forty-one years; in 1900 it was forty-nine years; and today it is about seventy years. In effect, thirty-five years of productive human life have been added. In this same period the world's population has grown from five hundred million to approximately four billion. Infant mortality has declined dramatically, in the developed countries from 400 per thousand to less than 1 per thousand. Whole categories of disease have all but disappeared. The major diseases of malnutrition are almost gone. The great epidemics are diseases of the past. Tuberculosis, which caused more than 300,000 deaths in the United States in 1900, killed only 5,000 people in 1970. This corresponds to a decline in the death rate from 113 per 100,000 to 3 per 100,000. The childhood diseases have declined markedly. The impairments caused by human disease are disappearing from the population. Today, 97.5 percent of the children born in the United States survive to adult life. But much of this has been accomplished, we must admit, by disturbing the balance of nature.

In all parts of the world there is a general statistical parallel between the degree of modernization of nations and the level of human health. Whether one measures modernization by per capita income, by level of literacy, by degree of urbanization, or by installed electric power capacity, the more modern the nation the greater the life expectancy of its population, and the lower the levels of morbidity. There are certain diseases which are much more common in modern industrial societies, as we have seen. Nevertheless, the frequency of impairments is much lower in modern societies, and the overall burden of chronic and acute illness in their populations is much lower also. Contrariwise, in preindustrial societies, where life expectancy is low, illness and death are phenomena of infancy and childhood, and they are produced by infection, malnutrition, and trauma.

Diseases of Modern Societies. In modern societies, where most of the people live well beyond the age of fifty, illness is a phenomenon of old age, and the outstanding causes of death are cardiovascular disease and cancer. In most of their forms, these two seem to be truly environmental diseases. The conditions of life in modern societies facilitate their occurrence, just as the conditions of life in preindustrial societies facilitate the occurrence of infectious diseases.

The other groups of diseases are special causes of morbidity in modern societies. These are, first, the addictions, and second, the disturbances of mood, thought, and behavior, along with some of the disorders of bodily function that often accompany them—that is, the neuroses and the psychosomatic diseases. The addictions are not new diseases. They create special problems in modern societies because technology has provided a host of new agents to which people can become addicted, and has made these new agents widely available. The use of wine, beer, and other fermented products (Figure 18.9) and substances such as hashish, opium, peyote, and coca leaves go far back, even into prehistory. All of these agents have the capacity to produce transient feelings of euphoria and elation, and people who use them repeatedly may become dependent on them.

Addictions became much more pervasive problems after the technological advances of the last three hundred years provided abundant supplies of distilled hard liquors, manufactured drugs such as heroin, and a variety of synthetic agents such as barbiturates and LSD, along with the marketing and transportation facilities to make them widely available. At the present time so many people throughout the world are addicted to one or another of these agents that they are a major cause of morbidity and mortality. They are also directly involved in a large proportion of accidental deaths. For example, at least 50 percent of fatal automobile ac-

18.9 Subano men (Philippines) drinking a fermented drink. (courtesy Peabody Museum, Harvard University)

cidents involve alcohol. Addictions are also the major disruption of families. This again is particularly true of alcohol.

A growing body of scientific opinion holds that tobacco is a truly addicting substance when it is smoked. Here again it was technology—the development of the cheap and convenient cigarette, long after tobacco was first brought to Europe—which made it possible for smoking to become widespread throughout the world. Cigarette smoking plays a role in forms of cancer other than lung cancer, notably cancer of the bladder, and it may have an effect upon coronary heart disease. However, its most important role in human health has yet to be widely appreciated by the public. Cigarette smoking and not air pollution is probably the primary agent responsible for the rising incidence of disability and death from emphysema in modern societies.

The prominence of minor disorders of mood, thought, and behavior may be related to the pattern of life in modern societies but it is difficult to be sure. The evidence is not adequate, and the disorders themselves are difficult to define and to classify. Up to 150 years ago, most people lived in a rural or semirural environment. There was little effective artificial lighting, and no really rapid means of transportation or communication. It is generally believed that the tempo of life in Western societies in those days was more closely geared to the daily biological rhythms of men and nature than it is today, when more and more men live in cities. A great many people follow a schedule of activities which goes on all day and part of the night, winter and summer. These activities tend to be scheduled with deadlines to be met, and tasks or responsibilities to be fulfilled. There is much contact of people with other people. There is continual alertness and arousal. There is rapid transmission of information, to the point of overload. There is difficulty in resolving promptly some of the conflicts and challenges with which a person is repeatedly faced. It is possible that this pattern of life may be responsible for the widespread symptoms of anxiety and tension, the sleep disturbances, the chronic fatigue, and some of the disorders of bodily function which are so common in modern societies. And it may also be, in part, responsible for some of the prevalence of peptic ulcer, of minor depressive symptoms, and of obsessive-compulsive symptoms. It is estimated that one man in ten has an ulcer, and at least that proportion has some behavioral or psychiatric disorder. In fact, data from the National Health Survey in 1960–1962, based on positive responses to twelve questions about psychological distress, suggest that one-fifth of the U.S. population had experienced an impending or actual nervous breakdown. On the other hand, we have already seen that the few cross-cultural studies that have been made show few differences in mental disorders from primitive to civilized societies. The same pattern of life may play some role in hypertension and coronary heart disease, but it does not appear to be the primary factor in either of these diseases.

Pollution and Its Effects

What about pollution? Ours has been called "the effluent society." From the point of view of the person who is affected, pollutants might be defined as potentially damaging biological, chemical, or physical agents which one inhales with the air that one breathes or ingests with one's food and water. Pollution is not a new phenomenon. It has been a problem for man as long as the species has been around. Our ancestors, like many primitive people today, drank polluted water, ate putrid and grossly contaminated food, and inhaled all sorts of pollens and dusts. Smoke-filled huts, tents, and houses have been human habitations back into the Paleolithic era (Figure 18.10). Among natives of New Guinea, lung disease is prevalent, due to allergy to the thatch used in making roofs. The lungs of Egyptian mummies have been found to be full of carbon particles. In London, several hundred years ago, Parliament passed a law regulating the use of coal fires, which were causing an intolerable smog.

So far as human health is concerned, pollution of air, food, and water by microbial agents has always been the most important form of pollution. This has been markedly reduced during the past 150 years. The water we drink in the United States today is probably more free from infectious agents than it ever was in the past, and the food is less likely to be spoiled or contaminated. Nevertheless, infections spread by air-borne droplets or dust are still by all odds the most common types of human disease which pollution causes.

Artificial Pollutants. What worries us today is chemical pollution—the introduction into the human system of those biologically unfamiliar chemical and physical agents which, acting in small amounts and over long periods of time, may lead to cancer, genetic abnormalities, or other serious diseases. There are many sources of such pollutants. It is estimated that during the course of a year, each American consumes from three to five pounds of substances which are added to food to preserve it,

18.10 Pincevent, a reindeer hunter's campsite south of Paris; over twelve thousand years old. Three fireplaces have been uncovered in what archaeologists believe to have been a tent habitation. Note the great scatter of animal bones within the living area. (courtesy Peter Timms)

to enhance its taste, or to facilitate its preparation. It is only fair to state that these substances cause relatively little disease, and that by aiding in the preservation, transportation, and processing of food, they make it cheaper and more available to many people. Their net effect on health might well be beneficial.

Further estimates are that each year the average American consumes from a quarter to a half-pound of pharmaceutical agents and drugs. The great majority of these medications are pain relievers, tranquilizers, and symptomatic medications designed to relieve feelings of fatigue, sleep disturbances, anxiety, headache, or functional disorders of various bodily systems.

It has been estimated that in the course of each year the average city dweller inhales forty pounds of carbon monoxide, five pounds of sulfur dioxide, half a pound of hydrocarbons, and half an ounce of nitrous oxide with the air he breathes (Figure 18.11). If a man smokes one pack of unfiltered cigarettes a day, during the course of a year he will inhale a half-pound of carbon monoxide, over a quarter-pound of tar, and about a quarter-ounce of other potentially harmful substances. The wonder is, not that these substances cause so much disease, but that they cause so little.

Pollution and Disease. Clear-cut diseases caused by air and water pollution occur most often among people with special exposures, such as asbestos workers, uranium miners, and textile workers. Up to now, community-level air pollution has caused clear-cut disease in large numbers of people only in those areas where unusually high concentrations of pollutants have occurred over relatively short periods of time. The excess deaths that have occurred under conditions of this sort have been chiefly

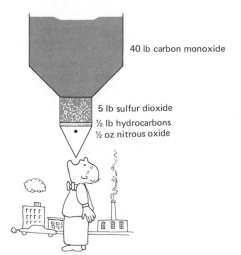

40 lb carbon monoxide

5 lb sulfur dioxide
½ lb hydrocarbons
½ oz nitrous oxide

18.11 The yearly amounts of artificially produced, and potentially harmful, substances ingested by urban American dwellers.

among people who already have lung or heart disease. This was the experience, for example, in the famous Donora, Pennsylvania, and London smog deaths. At normal levels in British, Canadian and American cities, there is no effect of air pollution on health. (This statement can be documented from testimony to Congress.)

For some perspective on air pollution, consider the problem of atmospheric lead, which has risen steadily since the introduction, in 1923, of tetraethyl lead into gasoline. A recent study of the lead content of human hair compared samples of hair obtained in the United States in 1871 and 1971. Which concentration was higher? There was ten times as much lead in human hair in 1871 as in 1971. A similar result was obtained from analysis of the lead in bones. Sources of lead that have been largely eliminated are those due to the collection of rainwater from lead roofing, storage of liquids in leaded jugs, and the use of lead in improperly glazed earthenware, in leaded paints, and in cosmetics.

As for mercury contamination of fish, it is known that swordfish caught many years ago were found to have more mercury than currently caught swordfish, which were hastily banned from sale by the Food and Drug Administration. We should be vigilant but rational as well.

Most of the serious chronic disease that is caused by polluted air is caused by self-pollution—that is, by smoking cigarettes. The pollutants in cigarette smoke are from ten to twenty thousand times as concentrated as those in the air of New York City. The same may be said of the effects of water pollution and of food additives. Substances such as detergents in drinking water, or cyclamates in soft drinks, have caused little or no human illness up to now. We are concerned chiefly about their potential for causing illness: the long-term, low-level effects that might not turn up for many years or even until the next generation. This is also the case with regard to pesticides and herbicides. Pesticide poisoning in humans has occurred mostly among farmers, plant sprayers, and other specially exposed people, or accidentally in infants and children. Up to now these agents have caused little or no detectable illness in the general human population. So far as human health goes, there is less concern about what these substances have done in the past than about what they may do in the future.

The two outstanding variables that are most closely associated with the general health of people in the United States today are income and age. Illness is more frequent among people with low incomes and among older people. In the United States today, even the poor are well-to-do by world standards. Without claiming that this is the best of all possible worlds, let us look at some hard facts provided by the Census Bureau after a national survey in mid-1971. The median income—that which half of all families are above, and half below—was $10,300. Of all families, 99.1 percent had refrigerators and stoves; 95.3 percent had at least one television set; 80 percent had at least one car; 75 percent had washing machines; 60 percent owned their own homes. So poverty, while real enough, is small in percentage terms and small in magnitude by world standards (Figures 18.12 and 18.13). All Americans, rich and poor alike, exhibit the patterns of disease most common in modern societies. Nevertheless, the differences between the health of the poor and the health of the well-to-do have the pattern of differences between the health of people in preindustrial societies and the health of people in modern societies. The poor have a higher infant mortality rate, a greater level of morbidity or illness, a greater level of impairments throughout life, and a lower life

18.12 Disposing of junked automobiles presents a considerable problem in the United States. (courtesy Richard H. Wagner)

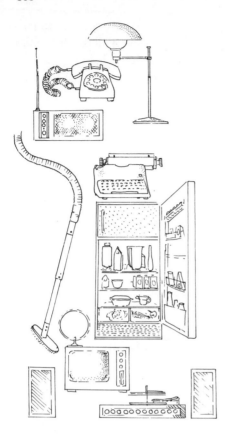

18.13 Material wealth in the Western world. A college student either owns or has access to many of these objects. Few are considered luxuries. Most use energy.

expectancy. This is partly attributable to higher levels of infection, of injury, and of malnutrition as well as to inadequate medical care. By contrast, high-income groups in our society have a high level of health wherever they live; whether in high density urban areas, in suburbs, or in rural areas. Their relative advantage seems to extend even to the uppermost ends of the scale. Managers and executives of corporations are healthier than workmen, foremen, and clerks. Men in *Who's Who* outlive their age-matched contemporaries. The patterns of life of members of high-income groups are characterized by high levels of sanitation, good nutrition, relative protection from trauma, high levels of care for mother, infants, and children, high levels of education, and, of course, high levels of consumption of goods and services, including medical care. Since these patterns of life are often concentrated in well-to-do suburban areas, some of the suburbs of our cities rank among the healthiest places to live.

The relation of age to health in modern societies is to a very large degree a reflection of the effect of modern civilization on patterns of illness. People in our society live a long time and, in general, do not become ill until they are old.

In sum, the following are the effects on health of the development of urban living. Men's lives have been prolonged and they have been freed from many diseases and impairments. New diseases have come forward to replace old diseases, but by and large, these new diseases are diseases of old age and diseases of affluence. New patterns of life seem to be creating many new symptomatic disorders, but not a great deal of mortality. The effects of new chemical agents—pollution if you like—are causing rather little human disease at the present time, but their potential for causing disease in the future is a matter of concern.

Effects on the Ecosystem. The implication of this is clear and important. The development of human civilization up to now has been a threat to the ecosystem, but it has been a boon to man. Men have achieved their present state of health and material comfort not by preserving their relation to the natural system, but by deliberately altering their relation to it. The farther men have moved from their original dependence on natural systems, the healthier and longer-lived they have become.

Now all this is not to overlook the threat to the ecosystem. We have become aware of the danger, and are concerned about it. We have been very hard upon the worldwide ecosystem that supports us. We have exterminated many species of animals and have altered many others. Our agriculture and the grazing of our herds have helped to turn the New East and North Africa into deserts and very nearly converted much of the American plains into a desert also. We have cut down the forest, dug up the mountains, and tainted the oceans with oil. We have turned the rivers into sewers and the skies into a chimney. There is a remote possibility that we may start disastrous earthquakes, impair the ability of plant life in the ocean to produce oxygen, or create a new Ice Age. All of this would threaten catastrophes of such appalling magnitude that the health or ill health of a few human beings becomes negligible beside them. But even apart from catastrophes, we cannot continue to spend capital, in the form of nonrenewable resources, at our present rate and still leave something for our children, our grandchildren, and the generations to come. We are still ultimately dependent on the worldwide ecosystem. If it goes down the drain, we all go with it. We can have no priority greater than that of preserving it. Yet when we begin to make decisions about specific moves related to the human environment, our dilemma becomes apparent. Human costs may be involved in any move we wish to make, and the implications of these can be important. We shall discuss this serious dilemma at some length at the end of the book.

Table 18.1. URBAN AND RURAL MORTALITY COMPARED

U.S., 1959–61 CHILDHOOD MORTALITY
RATES PER 100,000 POPULATION

All Cases Aged 5–14 Years

	Metropolitan counties	Nonmetropolitan counties
White male	48.6	59.7
White female	32.3	37.6
Nonwhite male	67.5	82.3
Nonwhite female	45.5	60.9

Urbanization and Vital Statistics. Concerning the city, we can all enumerate its characteristics: crowding, noise, odors, dirt, air pollution, traffic congestion, rising crime rates, coexistence of wealth and poverty, of architectural grandeur and squalor—a totally man-made environment, an asphalt jungle. How can health in the city be good? And yet it is, by our conventional measures of mortality and morbidity. Up to 1950, death rates for all causes in the United States were higher in cities, but by 1960 the ratio was reversed. Rural rates became higher than urban, and since 1960, the ratio of rural urban deaths has been steadily increasing. Thus, paradoxically, even though cities have been increasing in size since 1940, death rates have fallen more rapidly in these crowded circumstances than in the more sparsely populated rural areas. Part of this phenomenon may be due to the improved medical care and sanitation in the cities and part to the migration of younger people to the cities leaving an older, more susceptible population behind in the rural areas. These processes, it could be argued, might overwhelm or obscure the harmful effects of city life. But these can only be partial explanations for this reversal in the rural-urban health ratios. Look at the data shown in Table 18.1. The migration hypothesis cannot explain the excess mortality rates in rural children over urban children, both black and white, male and female. Other data from the United States show higher rates of infant mortality in rural areas, and higher rates of rejection of draftees on health grounds.

Data from other parts of the world tend to confirm this seeming paradox. For example, according to René Dubos, despite the fact that Hong Kong and Holland are among the most crowded areas in the world, they enjoy some of the highest levels of physical and mental health in the world. Even more convincing are the data from Britain, where in 1961 the age-standardized mortality ratios for all causes of death were as follows:

	Males	*Females*
Urban areas, population 100,000+	101	98
Urban areas, population 50,000–99,999	91	90
Urban areas, population under 50,000	104	105
Rural areas	91	98

In other words, there is no significant trend in mortality rates between city and country, or between different sizes of cities.

To take a specific cause of illness and death, tuberculosis has been used as an excellent example of a disease which, following industrialization and increased crowding into cities, showed a marked increase in rates. Not so well recognized, however, is that in all countries for which data are available, tuberculosis programs cannot account for the following reversal in trends, at least in the beginning. The decrease of tuberculosis in Britain and the United States, for example, started fifty to one hundred years before any useful antituberculosis drugs were discovered, and several decades before any organized antituberculosis programs were initiated.

Furthermore, in some recent studies it has been found that, contrary to prevailing theory, tuberculosis does not necessarily occur under crowded conditions. Under some circumstances it occurs more frequently in people who are socially isolated. In one British study, while there was a strong social class gradient in the prevalence of tuberculosis, no relationship was found between prevalence and crowding. In fact, lodgers who were living alone had a tuberculosis rate some three to four times higher than family members, even though the lodgers were, by definition, living in uncrowded conditions. Similar results were found in a study in the United States in which it was found that tuberculosis was occurring most frequently in people living alone in a single room and not in those living under the most crowded conditions. Such data do not necessarily refute the role of crowding in changing susceptibility to disease, but they do provide clues which may require a change in our thinking about the processes through which crowding can influence health and the circumstances under which this occurs.

SUGGESTED READINGS

Banfield, Edward C. 1970. *The unheavenly city. The nature and future of our urban crisis.* Boston: Little Brown.

Dubos, René. 1969. *Man adapting.* 4th ed. New Haven: Yale University Press.

Ehrlich, Paul, and Ehrlich, Anne. 1972. *Population, Resources, Environment*, 2d ed. San Francisco: Freeman.

19
CHANGING NUMBERS OF MANKIND
By Roger Revelle and W. W. Howells

The Recent Population Explosion

It is a truism to say that we live in a time of unprecedented change and, even more striking, of very rapid rates of change. In a few short generations we are consuming the coal, petroleum, natural gas, and lignite that accumulated in the earth over the past five hundred million years. Within our own generation we have discovered and tamed the energy in the atomic nucleus and in so doing we have attained the terrible power to destroy all civilizations, and most human life, in a day. The rate at which we are gaining knowledge of the universe and of living things is so fast that no man can keep up with more than a small fraction of our exploding knowledge.

One of the unprecedented changes of our time—in some ways the most fundamental change—is the increase in the numbers of human beings; the world's population is believed to be growing at about 2 percent per year. At first thought this does not appear to be a very rapid rate, compared to interest on investments of 6 to 12 percent, or inflation of 10 to 20 percent per year in many countries. But a 2 percent rate of population growth means that human numbers would double every thirty-five

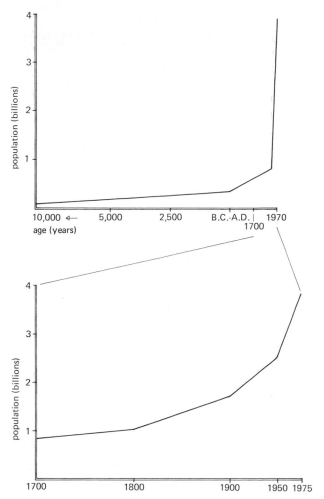

19.1 The world's population. Most of the world's population growth has taken place since the development of agriculture ten thousand years ago, or in less than 1 percent of the time man has been on earth. Various authorities have estimated that in the million years which preceded agriculture, the world's population grew to no more than 8 million, but only eight thousand years later in A.D. 1, the population was almost 300 million. The billion mark was reached by 1800. By 1900, this figure had almost doubled. Between 1950 and 1974, the rate of growth more than doubled that of the first half of the century. Today, the world population is estimated to be 3.9 billion. More than 80 percent of this human increase took place in less than 0.02 percent of man's history. (Redrawn from Coale, Ansley, J. 1974. The history of the human population. *Scientific American*. 231:42.)

years, and increase tenfold in about one hundred twenty years. Such a rate of world population growth has never existed before, during the hundreds of thousands of years of human history (Figure 19.1). It seems highly unlikely that it can continue for very long in the future. These statements are easily demonstrated if we do a little arithmetic.

Suppose, for example, that world population had been increasing at a rate of 2 percent per year ever since the battle of Hastings in 1066, reaching the present figure of around four billion in 1976. Three hundred fifty years ago, in 1626, there would have been less than four million people; seven hundred years ago, in 1276, the population of the entire earth would have been less than four thousand, and two hundred and ten years before that, in 1066, there would have been only about fifty people on earth. Yet, unless someone has been pulling our legs, there were a good many thousand soldiers on each side when William the Bastard defeated Harold the Saxon king at Hastings, and became the Conqueror of England. All the evidence indicates that several hundred million people were living on the earth at that time and that this has been true for many centuries in the past. Rates of population growth throughout most of the lifetime of our species could not have been more than a very small fraction of 1 percent per year (Figure 19.2 and 19.3).

If we look into the future, and assume that the present rate of population growth of about 2 percent a year were to continue unchanged, one thousand years from now there would be more than a billion billion human beings; one for every square centimeter of land area on earth. Of course it would be possible for the earth to contain this number of people, by piling them on top of each other. If the entire land surface were covered with buildings ten thousand stories high, each person would have one square meter of living space. Or alternatively, through the miracles of genetic engineering, we might produce a race only a few inches tall. They might be able to survive under such crowded conditions, but from our present standpoint this seems neither probable nor desirable. Rates of human population growth must slow down and eventually become zero.

Indeed, if human beings are to continue to live on this planet with some prospect of happiness for the majority of mankind, population growth will need to slow down very nearly to zero within the next two or three generations. This seems evident when we consider the earth's potential for providing food. Even supposing that all the actual and potential farmland on earth, which covers an area of about two and a half billion hectares, were to be cultivated at the level of the best present agricultural technology, and two or three crops were grown each year wherever temperatures and water supplies permitted, the food produced would be sufficient for only about forty billion people—ten times the present number. The amount of energy required to produce this food and distribute it to people would be enormous, about equal to the total energy in coal, oil,

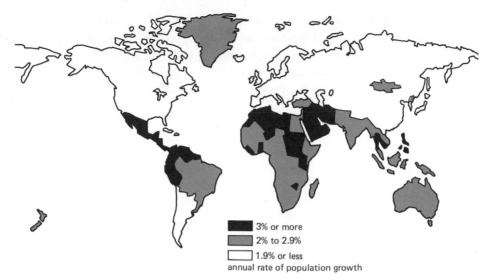

19.2 World population growth pattern through the 1970s. (SOURCE: *Population Program Assistance*. 1972. The Agency for International Development.)

natural gas, and water power used by the world's society at the present time. But such a high level of agricultural technology would be possible only if there were worldwide industrialization, which would require many times the world's present energy consumption. Without a series of technological breakthroughs and a very high level of capital investment, no known energy resources could provide this energy for very long. Yet, at present rates of growth, a population of forty billion would be reached before the end of the next century.

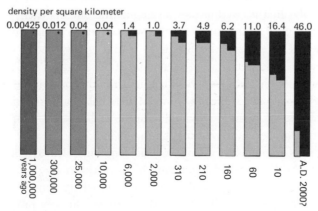

19.3 World population: Human density per square kilometer.

It is this kind of arithmetical thinking, plus the observed fact that for the majority of mankind who live in the poor, so-called less developed, countries, food production is barely keeping up with population growth and in large areas is lagging behind, that has revived the old ideas of Thomas Robert Malthus, the parson turned economist.

Malthus
the Pessimist

In 1798 Malthus published a slim volume entitled *An Essay on Population*, one of the most influential books ever written. The essence of his argument was this:

The "passion between the sexes" will undoubtedly continue to be as strong in the future as it has in the past.

Therefore human beings will continue to have many children. (Malthus was unaware of the technical possibilities for contraception that would emerge during the nineteenth century, and besides he thought that contraception was a vice.)

At the same time the most fundamental need of human beings is food.

From these propositions Malthus deduced his Principle of Population which was simply that human populations, like those of other animals, will always increase up to limits set by the food supply. Human populations tend to increase geometrically in accordance with the laws of compound interest, whereas Malthus thought that food supplies tend to increase arithmetically, that is, by more or less equal increments each year. Consequently, at most times and places throughout history, large numbers of human beings have barely been able to obtain enough food. The size of the population has been limited by malnutrition, disease, and war, or, as Malthus said, by "misery and vice." Whenever the miseries and vices of human beings are not sufficient to maintain a balance of population with food supply then "gigantic inevitable famine stalks in the rear and with one mighty blow levels the numbers of mankind."

Malthus's friend, the great classical economist David Ricardo, deduced from the Principle of Population his own Iron Law of Wages—The incomes of poor laborers can never rise for very long above the level of bare subsistence, because if they do the poor will simply produce more children, and the pressure of excess numbers of people wanting jobs will drive wages back down again to subsistence level. This is why economics used to be called "the dismal science."

Malthus recommended that the Poor Laws of England, which dated back to the time of Queen Elizabeth I, should be abolished. As long as each parish was obliged to support its own poor, these poor people were able to produce more children and the total amount of misery was thereby increased. He recommended instead that workhouses for paupers should be established throughout England. The sexes would be separated by incarcerating men and women in separate workhouses. These workhouses would be such unattractive places that the poor would work very hard to stay out of them, thereby at least increasing labor productivity even while the numbers of poor people increased.

The *Essay on Population* encountered a mixed reception in England. The Tories were delighted, the Liberals were appalled. Malthus was violently criticized by the very people whose opinions he valued.

The *Essay* had been written without using much actual data. It was largely a deductive exercise, based on Malthus's three quite reasonable-seeming propositions. Malthus decided that before writing anything more on the subject he had better find out what was actually happening to the populations of different European countries, and for the next five years he made a grand tour of Europe, gathering data and observations. He found that the people of Switzerland thought their population was declining because the number of baptisms recorded in the parish registers was diminishing each year. Because population change depends on the difference between births and deaths, these data were clearly inadequate by themselves to demonstrate what was happening to the Swiss population. By examining the Swiss parish registers, Malthus found that the number of funerals had also diminished. Further investigation showed that fathers and mothers in Swiss farm families were living longer on the average than in former times. Without a farm of their own to enable them to support a family, Swiss young people were postponing marriage until their parents died and they could inherit the family farm; hence the decline in the numbers of births. The Swiss population was still growing, however, since the numbers of births exceeded the numbers of deaths. Elsewhere in Europe, Malthus discovered similar behavior patterns among the mass of the people, and he returned to England convinced that human beings can and will limit the numbers of their children by practicing what he called "moral restraint," that is, postponing marriage. Moral restraint was a "preventive check" on population growth which might be as effective as the "positive check" of famine and disease. He wrote a new edition of the essay on population which was many times longer than the first and much less simplistic in its conclusions. Unfortunately, like many famous books, even the first edition of the *Essay on Popuation* is seldom read, let alone the long and complicated second and subsequent editions. Most of the modern neo-Malthusians have learned about Malthus at third or fourth hand and they are familiar only with the Principle of Population, which was stated so eloquently and inaccurately in the first edition.

"The Demographic Transition"

During the nineteenth century and the first decades of the twentieth century the populations of Europe and of people of European stock in the New World grew very rapidly. Yet at the same time it was evident to everyone that the conditions of life of most members of these populations were steadily improving. Malthus's gloomy views, while not forgotten, slipped into the background. It is only during the last two or three decades, with the enormous growth of populations in the poor, so-called less developed countries, that the ideas of inevitable growth of human populations up to limits set by the means of subsistence have again become widespread. During these same decades we have learned a good deal about the actual behavior of human populations living in different environments. Contrary to the views of the neo-Malthusians, many population scientists now believe, just as Malthus himself concluded in the later editions of his essay, that human population growth up to levels set by the means of bare subsistence is far from inevitable and in fact seldom occurs.

The idea of birth control (other than Malthus's "moral restraint") had appeared early, and had its enlightened pioneers during the nineteenth century, but also massive resistance. Just a hundred years ago such pioneers in England were being vigorously prosecuted under antipornography laws, for corrupting the young. Anticontraception statutes of one kind or another remained solid in the United States through the first half of the twentieth century. In Massachusetts they were installed by Yankee cousins of the English protectors of the young, although after they had passed the torch to the Catholic Church their own descendants labored to repeal them.

Legalities and religion, however, did little to stem the tide once it turned, and birth rates in the developed countries everywhere began to fall. Religion was less of a wall than might have been expected, with Catholics following Protestant trends, although some percentage points behind. It is now strange to realize that in the period before World War II some European countries, including France, Germany, and Sweden, had such declining birth rates (although with lower death rates there was still an increase in the total population) that governments became worried for the future and began offering inducements for larger families. The Soviet Union soon after the 1917 revolution provided free abortion, but stopped it in 1936, although it was installed again in 1954. In the United States, birth rates reached a low point in the 1930s, and the trends of that time were responsible for some of the gross underestimates of present population sizes here. After the war, with the well-known baby boom, our birth rate went up, but has since resumed its decline, The average number of children born to one thousand women during their lifetime (the total fertility rate, a computed figure, to correct for variations in births to women of different ages) was 3,459 for the years 1960 to 1964 and 1,896 for 1973, a dramatic drop of 45 percent. (The percentage drop was greater among Catholics than among non-Catholics, and about the same among nonwhites as among whites. These and various older figures show that our population has been getting more homogeneous in the fertility of its different segments; earlier in the century, urban dwellers, and those of higher educational levels, had much lower birth rates than their opposites; the differences are still there, but decidedly diminished.)

Such reduced birth rates do not, however, mean that our population does not still increase. Here, as in some other advanced countries like Canada and New Zealand, death rates have also been brought so low that the surplus of births continues to be considerable. However, we are pointing toward a gradual stability. We in the so-called developed countries have gone through a long cycle, from ancient high birth and death rates to strongly controlled low birth and death rates. A most striking fact is that in this country unplanned or unwanted births have recently fallen distinctly faster than fertility as a whole. Conscious choice by individual families is thus approaching complete determination of population numbers in the United States, in this final stage.

This historical phenomenon, in which a transition has occurred within less than two centuries from high and variable death rates and high birth rates at the beginning, to low and relatively constant death rates and low but somewhat variable birth rates at the present time (Figure 19.4), is called by population scientists the "demographic transition."

The cycle took off from the situation in early societies, those of hunters and gatherers, and simple gardeners. It might have discomfited Malthus, but anthropologists find that these precivilized folk today give

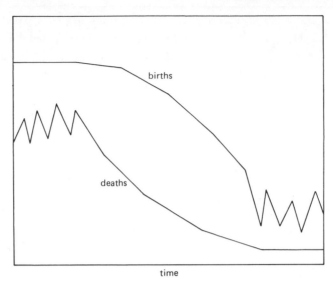

births

deaths

time

19.4 The demographic transition illustrates the script which has been followed by industrialized nations of the West and, according to the more optimistic of demographers, will be followed by those of the third world. Advances in medicine and public health bring about a decline in the death rate (bottom curve). Later, with the modernization of the society, children are no longer perceived in terms of the economic assistance they afford. Note that, before the transition, the birth rate is constant but the death rate varies due to the vicissitudes of a life not touched by health service. After the transition, it is the death rate which is constant and the birth rate which varies. In the third world so far, only the first part of the script has been read. Death rates, while high, have declined, but birth rates remain high.

little evidence of stressing their resources by overbreeding; and their checks are not vice and misery but, as we shall see, fairly high death rates in the young and some deliberate restraints on numbers of children. Nomadic hunters and gatherers have natural limits on their food supply, and the problem of carrying very young children and giving them suitable food other than mother's milk leads them to space births as well as they can, backstopped sometimes by letting a new infant die when the previous one is too dependent. But both hunters and gardeners have a surprising amount of leisure time, and it can be shown that the latter can often extend their food supply by some extra effort.

Then came the cycle of change, the increasing rise in populations, and the explosion which is still going on in the peasant societies of the underdeveloped nations. How, after so long a period of very modest expansion of mankind, have there begun to be enormous numbers?

The arithmetic is simple: more people are being born, and they are living longer. But both of these facts need a little more examination. We may in fact first ask whether modern humanity has changed from its Stone Age ancestors, either in the possible numbers of children it can produce or in its ability to live a long time. The answer to both questions seems to be no.

The Biological Component

We might notice here a principle followed by demographers, which could be called *regardez la femme*, or "keep your eye on the female of the species." It is not sexist to recognize that she is the actual producer of children, and that she can produce only so many during her life. Therefore, statistics can be usefully analyzed and compared only if they are geared to what women, of different ages and in different places, are actually doing in this line. Of course, reproduction is affected by a male's infertility, absence in war, or death, but if he is present and fertile, his age is of no consequence, whereas that of a woman is. Even a man's state of health counts for little; it has been shown that physical exhaustion does not impair his immediate fertility.

Fecundity. Man is a slow but persistent breeder. Has the absolute number of children a female may bear changed over time? This is difficult to measure, because it is easily masked by changes in expressed *fertility* (which, when used as a technical term, means in English the actual number of children, or rate of their production, as against the biological potential, or *fecundity*). The number of children ever born to all women past the child-bearing age fell, in the United States, from over 5 in the middle of the last century to under 2.5 in the middle of this one, something obviously due to social, not biological factors. Among the Hutterites, a strict and isolated religious sect living in farming communities in the upper plains of the United States and Canada, the number of children ever born to married women has for decades averaged over 10, apparently the maximum of record. (In general records, occasional women are known to have given birth to as many as thirty children, but these are highly exceptional and have no real effect on the population.) The Hutterites marry at about twenty, abhor birth control, and avoid frivolities, but spend much money on proper medical care. Their total increase in a century since immigration has been phenomenal, but they are a wholly exceptional population. Sterile couples do exist these among them, but they are rare.

A careful analysis of married couples living in Indianapolis led to the conclusion that actual sterility (due either to husband or wife) was relative: about 10 percent of couples would have remained childless, about another 10 percent would have had a single child, another 10 percent two only, and so irregularly on. Apparent lowered fecundity, especially among city dwellers, has in the past given rise to rather bizzare explanations, such as stress of urban life, and even walking on pavements; these can be safely ignored. At the other end of the scale, among aboriginal peoples (but not necessarily "Stone Age" or primitive) figures are hard to establish, since women are apt to forget miscarriages, abortions, stillbirths, or infant deaths. However, one careful examination of Pueblo Indian women past child-bearing indicated that the average number of pregnancies had been about 9.4. (This includes miscarriages and stillbirths, and so does not compare with actual Hutterite births.) This study was done at a time when health care had not long been established for these Indians (i.e., the women questioned had probably had little benefit of it earlier, during their child-bearing years).

A biological feedback mechanism may also operate in both primitive hunting and gathering, and traditional agricultural societies. Rose Frisch has shown that women who do not have a sufficient ratio of fat to total body mass do not menstruate and presumably do not ovulate. Also, the age of menarche is delayed and perhaps the age of menopause is hastened when girls and women are too thin. Thus, when times are hard and women do not obtain enough to eat, their fecundity may be lowered. This may aid the population to come into balance with the food supply and to rise well above subsistence levels when environmental conditions improve.

Length of Life. In advanced countries the life expectancy of a newborn child is now over seventy years, following a long and steady rise. Many nonadvanced countries have expectancies falling below forty-five, while for aboriginal groups figures, like those for total offspring, are hard to make out, again because of no great interest as to personal age and the tendency to forget some who died young.

Does this mean life is really being extended? Every now and then some doctor promises that people will soon be living to 125 or 150; and there are those strange peoples, like the Hunza, where hale and hearty centenarians are alleged to be common. Are we really breaking old-age barriers? No. Many of us are indeed reaching the biblical promise of three score years and ten, and many are also going well beyond. But alleged cases of great age are seldom substantiated. No Civil War veterans, for whom records should be good, reached the age of 110. In the 1960 census of Hungary recorders reported sixty-seven people aged 100 or more, but a further check of these showed that only fourteen could be so proven; the oldest was a woman of 103 who died six months later. Ages seem to be highest where records are probably sloppiest. When, for censuses taken since World War II, we read that for every 100,000 of the population, the number of centenarians was 0.6 in England and Wales, 1.2 in Sweden, 0.7 in France, and 5.8 (!) in the United States, all advanced countries with good health care, while in Egypt it was 27.1, in Guatemala 23.5, in the Dominican Republic 54.1, and in Iraq 90.8, what should we think? We should think twice.

The statistical rising life expectancy reflects things other than great age. We know from tombstones and other records of Greece and Rome that a number of individuals lived to what we look on as old age, but the proportion of these was fewer. The majority of those who today would survive were struck down along the way by disasters we would now consider routinely avoidable accidents: typhoid fever, ear infections, operable cancers, to take examples from three kinds of recent medical conquest. Albert Damon could recall six occasions when modern medicine saved his life, and many of us, even if not medically trained, might be able to think of at least one such personal escape. If one knows his family history in the last century at all, he or she may realize that an undue number of great-great-uncles and aunts had untimely deaths.

Among simpler peoples things were surely worse. In the study of Pueblo Indian maternity referred to above, infant mortality was very high: one-quarter of the children died in their first year and altogether one-third before the end of the second year. With wretched sanitation, many mothers died of childbirth-related infections, with the usual result that the infant died also, from lack of care or of mother's milk. For most primitive populations, we can learn more from the dead than the living—their skeletons are better guides to age than their memories. Of Neanderthal men, living before 35,000 B.C., possibly one of those found is believed to have reached the age of fifty, and this held through the Upper Paleolithic of Europe—possibly one individual or two (different authorities) out of one hundred and two skeletons examined passed that age. Among the inhabitants of Indian Knoll, Kentucky, about 3000 B.C., people aged fifty or more were almost as rare as centenarians among ourselves. By the time of classical Greece and Rome figures were better: 13 percent for Greece, more for Rome. But in the United States today the figure is nearer 90 percent: no wonder that life expectancy has gone from thirty-eight in fourteenth century England to over seventy today. The very rapidity of the rise makes it evident that biology has not changed, and that mankind must have had the same possibility for long life from the beginning, although it was almost never realized. (Obviously, numbers of the living have increased from this cause alone, as the average individual survives through a much longer span, and whole age classes above fifty fill out greatly.)

The raw figures above mask a fact crucial to the population explosion of today. What is important is not so much how many die old as how many die young. For former populations, counts of the young are difficult from skeletons or early records—infant bones are more easily destroyed, and infant deaths neglected. If we take all those who died before maturity—twenty or twenty-one—combining infants, children, and adolescents, we have some approximate figures, for early communities, of their proportion among total deaths. A large Iroquois ossuary gave 57 percent, while the large Indian Knoll cemetery gave about 50 percent. Two Hungarian leading specialists, Georgyi Acsádi and Janoš Nemeskéri, find figures from about 40 percent to over 50 percent for Roman and early medieval cemeteries or epitaphs, probably the most careful figures. Others have reported: Middle Bronze Age of Greece (about the time of the Trojan War), 59 percent; Early Iron Age of Greece, 67 percent; tenth-century Czechoslovakia, 62 percent. There is a considerable margin for error in these and other figures, but we might accept that in a typical premodern community some 50 to 60 percent did not survive to marry. (The wastage may actually have been greater, by some signs, after the cycle of population rise had begun, leading to larger, more closely packed societies, than it was in the tribal communities preceding them.)

In an absolutely stable population (zero population growth) each woman must replace herself, no more, no less. That is, the average woman must have enough children so that, given the vicissitudes of survival, marriage, and reproduction, she will have had on the average one daughter who will do the same in her turn. (This is *regardez la femme* in earnest; demographers call the above a "net reproduction ratio" of 1.00, just in balance.) Obviously, if chances of death are high, the required number of total children will be high.

Now consider a society with the grievous mortality rates we have just cited. Imagine a family of seven children. Let us say that four, or 57 percent, will not survive to reproduce. Of the three survivors, we may expect one or two to be daughters. Under the same conditions of mortality it is up to the surviving daughter or daughters to reproduce the seven from

which they survive, if numbers are to stay the same. It is not hard to see
that the population would increase only against odds and that rather simple methods of control, where desired, would be effective (no renewal of
intercourse before weaning the last child; crude abortive or contraceptive
measures, occasional infanticide.)

Death Control

Here we come to the essential fact in the population explosion of today,
which is going on mainly in peasant societies (Figures 19.2 and 19.5).
Present-day medicine, public health, and perhaps better nutrition, are
preserving a far greater proportion of the young into maturity and through
most of the reproductive period. This has already had its effect in advanced societies over the last century, but we have seen how in them,
birth rates have fallen along with death rates. It is in the underdeveloped
regions that broadly applicable measures such as malaria and smallpox
control and general sanitation have been having the great impact, sharply
cutting down infant deaths while births still remain high (Figure 19.6).

  Consider again the hypothetical family of an early or undeveloped society with its seven children, three or four of them girls. The one or two

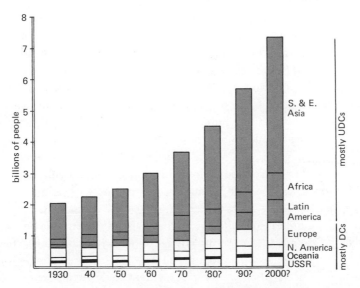

19.5 World population growth is shown for developed countries (DCs) and underdeveloped countries (UDCs). (Projections after 1970 are from U.N. estimates.)

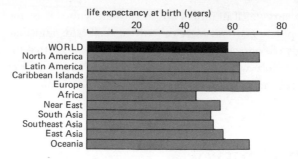

19.6 Life expectancy at birth based for the most part on data from 1973. (SOURCE: Social and Economic Statistics Administration, U.S. Department of Commerce, Bureau of the Census.)

previously surviving young women have a better chance of lasting through reproductive life, but that is a small matter. They are now joined by two of their sisters, no longer little tombstones but thriving mothers, reproducing at the customary high rates of their society. The sudden effect on total population numbers is not hard to see.

To estimate a rate of growth for the world as a whole is rather meaningless, and difficult to do as well. However, as previously stated, most demographers think this rate is about 2 percent a year at the moment, which would double the total population every thirty-five years. Here the differences from region to region are great, and important, even if the number of percentage points seems small. Growth rates for developed countries (which include Europe, North America, the Soviet Union, Japan, Australia, and New Zealand) are under 1 percent as a rule, while for Latin America they range up to 3 percent and beyond (Figure 19.2). Africa and Southeast Asia are also high. And here is another important point. The existing populations of South and Southeast Asia have become so large that small percentage increases translate into vast numbers of actual new human beings. It goes without saying that these are the countries least able to afford such new numbers. Our own population is still going up, and will do so for several decades even though actual births have dropped below the replacement level, that is, the level at which the average woman is just reproducing herself. Such a level is not the same as zero population growth because the still relatively large numbers of children who are already born will be entering and going through the reproductive ages for some time. This effect will hold in the large, less developed countries for a much longer period, because they are still far above replacement level, which is one reason why estimates of the future world population have been so great.

What will the future hold for these countries? We can hope that the demographic transition will follow the path traced out in the developed countries and that birth rates will come tumbling down to match the decline in death rates. But an alternative future is also possible and to many people it seems more likely. The populations of the poor countries may become stabilized in the future, not by a decline in birth rates but by a rise in death rates to their previous high level. This would be a ghastly outcome. With the modern technology of medicine and sanitation, most deaths would not be due, as they were in the past, to malaria and other infectious diseases but to starvation and malnutrition and the diseases of abject poverty. Most of the dying would be children because children are more often victims of malnutrition than adults.

Is there anything that can be done to make the first alternative— completion of the demographic transition—more likely than the second one? To answer this question we need to look at what is known about the causes of the demographic transition in the presently developed countries and to find also what has happened in those poor countries where the second stage of the demographic transition is now taking place. Various approaches, direct and indirect, are possible.

Birth Control

Everyone agrees the increase is serious and should be dealt with. Everyone agrees on death control also: disease and famine must be overcome, and even a Malthusian zealot who felt, "It's their fault; let them starve" would surely be shaken if it happened. In fact, everyone not prejudiced by special considerations is bound to recognize that the solution must be social, by limitation of births. There are violent disagreements over means, as opposed to ends, from abortion to Malthus's moral restraint. But that is another subject, and one not necessary to consider. For the fact is that if a people wish to limit their births, they can and will. We have already cited the advanced stage of limitation by choice reached by American married couples. To express it again slightly differently, between the first half of the 1960s and the second, the rate of frankly planned births remained the same, while unplanned births fell from a level about one-fifth above planned births to a level about one-fourth below them. And here is another kind of lesson. The computed average number of births per woman declined in Romania, with its permissive abortion law, from 3.1 in 1955 to 2.0 in 1962—below the approximate 2.1 level needed for replacement—and then to 1.9 in 1966. Romania reacted by banning abortions, and in a single year the rate shot up to 3.7. But it has been coming rapidly down again as the people reacted in turn by resorting to contraceptive methods and to illegal abortion.

The eventual choices controlling births are made by individuals and families, but social norms strongly affect the choices, so that it is whole classes and nations that manifest high or low increases. In a country like ours total numbers are little affected by the occasional family that gets photographed on Easter, with a brood stretching across the street. It is the millions of standard American families that choose between two or three or four children who set the trends, so that mode changes in ideals can thus shift birth rates rapidly, playing havoc with expert predictions over a decade or so.

The Means. Methods of limiting births are broadly of two kinds, which we could call technical versus behavioral. The first includes every sort of measure to frustrate pregnancy (or by abortion to end it), while permitting normal sexual activity. Such measures allow a high degree of control in developed societies today. Elsewhere, ignorance of them is a factor, but only one, in the continuance of high birth rates. Such efficient methods as the Pill may not only allow married couples exact control in planning but, by giving similar confidence to many single women, may also contribute to the second kind of control, the behavioral.

This relates to delays of marriage or of child-rearing. It is an important fact that women, many of whom are able to start bearing children in their middle teens, are most fecund, or likely to conceive, in their early twenties, and that this fecundity gradually declines (varying, of course, with individuals) from then on. Thus delay of marriage not only reduces the total span of a woman's exposure to conception but also removes the time of her peak fecundity from her reproductive life, progressively lengthening the odds against conception (assuming no preventive measures). It is a common story that some couples, putting off either a first child or an additional one, may find themselves disappointed in their expectations after all.

These are the actual tools of control. The root causes of choice are social and economic. Do people or nations choose to have fewer children because they are poor, or because they fear overpopulation? It is not so simple. Let us consider cases.

Who Is Doing What. The Irish, strict Roman Catholics, are probably the only nation fully to apply Malthus's moral restraint to help hold the population down. Apart from the painful emigrations following the potato famine in the middle of the last century, they have been faced, in a farming country, with problems of small land-holdings and possibly landless offspring. The key to fewer births has been late marriage and a very high degree of celibacy. In the 1950s, of women in the ages twenty to twenty-four, over 80 percent had not yet married, as compared with 32 percent in the United States; and more than a quarter of older Irish women had never married at all, or at least not in their reproductive years. Spots of Austria have even higher rates of spinsterhood than Ireland. In certain villages in Hungary, another Catholic country, in the late nineteenth century, people married young but held down the number of children to one if possible, so that as with ancient royal families a marriage could inherit and merge two separate farms. The method was evidently abortion, which still has a very high rate in the country.

Japan and China lack religious compunctions about birth control, and have been more forthright about limiting increases. The Japanese are a homogeneous nation, traditionally given to adopting what strikes them as a good idea. After World War II, contrary to their expectations of a few years earlier, they found themselves confined to the home islands, without prospects of emigration or colonizing. By wide acceptance first of abortion and more lately of contraception, birth rates have been brought very low, although death rates have also dropped to low levels so that some increase continues. In the Peoples Republic of China things are more frankly bureaucratic. The massive effort of one generation has at last allowed the nation to feed itself adequately after centuries of famines. With perhaps nine hundred million people (no one, not even the Chinese government, knows within one hundred million), strong limitation of births is now government policy, and government policy gets attention. Severe official and peer pressure is exerted on the young to marry late, and equally severe pressure to limit a couple's children to two, by contraception, abortion, and sterilization. At the same time, universal medical care of fair quality may have spread a new view of chances of the survival of children, thus perhaps lessening the fear of parents that their children will not survive to take care of them in their old age. At least in cities, birth rates have probably come down sharply in China.

India has much better statistics than China, but the statistics are not very encouraging. Each year, India adds about thirteen millions to her numbers, more than the total population of many smaller members of the United Nations. The most strenuous efforts to increase food production have barely kept ahead of the increase of population. The Indian government has made attempts to promote birth control ever since 1948, and recently these efforts have become much stronger. One state, Maharashtra, is drawing up legislation compelling sterilization for couples already having two or three children; others have offered incentive payments for those accepting sterilization; and the federal government is providing disincentives, including loss of jobs for government workers having more than two or three children who refuse to be sterilized. The reaction of the people to such a draconian approach remains to be seen. With many religions, languages, and ethnic groups, India is a complex country, and what might be successfully introduced in one area does not necessarily spread to others. More important, the numbers of children are not the result of accidental, thoughtless breeding. A poor farmer or laborer may see his children as his wealth, having no other: "They don't cost much, and when they get old enough to work they will bring in money. And when I am old they will take care of me." This is a quote from a father of eight (*New York Times*, May 30, 1976). Even though more children now survive and even if the government were able to assure care of the aged, it is a question whether such views would change quickly. And only changing views, not cheap contraceptives, are likely to bring birth rates sharply down. Most poor peasants and landless laborers in India lack the most important spur to birth control; having fewer children is not likely to improve the parents' position, nor to give greater opportunities—of education or social climbing—to the children they have. Nevertheless, the birth rate in India has dropped by at least 15 percent since 1950.

And more change can come. Under British administration of Fiji, numbers of Indians immigrated in the last century and rapidly moved toward becoming more numerous than the native Fijians. But not long ago the Indians shook off their traditional attitudes toward family size and their birth rate fell to that of the indigenous people, so that by 1971 the two groups were of equal size in the total population of half a million.

Vigorous programs to spread birth control have been undertaken in many other less developed countries. In two of these, Indonesia and Thailand, results just began to be seen in the early 1970s. Indonesia is one of the world's large nations, but also a complicated one, of many islands and ethnic groups; it has a very low per capita income, and high fertility; its strong family planning program is organized in each village around the local headman and his wife, and in some areas, it seems to be remarkably effective. In highly rural Bali, 55 percent of married women are said to have been practicing contraception by the second quarter of 1975, and in East Java, the figure was 32 percent. In Thailand there was an increase, in the three years before 1972–1973, of married women using birth control methods to almost double, to 26.3 percent. The largest proportion of users was among older women, but the greatest proportional increase was among the younger, perhaps an important indication. From 1950 to 1974 the Thai birth rate is believed to have dropped by 27.5 percent, from forty-seven per thousand to thirty-four per thousand.

In a dozen or so other relatively poor countries, including Korea, Taiwan, Singapore, Hong Kong, Malaysia, Sri Lanka, Mauritius, Trinidad and Tobago, Costa Rica, Barbados, Puerto Rico, Chile, and perhaps Egypt, birth rates have recently been declining rapidly, more so than was ever seen during the period of demographic transition in the rich countries. The data for these suggest forcibly that factors other than socially or politically fostered birth control may be having significant effects—economic and cultural factors.

Other Countries, Other Causes

Such factors may be seen in the passage of developed, urbanized nations through the demographic transition. As we have seen, in peasant societies children help in farming and do not cost much, either directly or in benefits lost; also, large numbers were necessary to ensure family survival. The status of women—prestige in the community and personal happiness—depended on their ability to produce children, especially sons.

With the coming of the industrial and agricultural revolutions in the countries that are now rich, all these conditions changed. The needs for farm labor rapidly diminished as the abilities of individual farmers to produce more food increased. People moved in vast numbers from the countryside to the cities, both because of unemployment in the country and because of the new opportunities for employment in industry and in urban services. For city dwellers, children are not the unmixed blessings they were in peasant families. Children contribute little to the family livelihood, especially when rising standards of living result in laws against child labor, and children are expensive in the city. Not only must they be clothed and fed, they must be educated and amused. In a city it is possible for women to find outside employment, but they cannot take a job if they must stay home to take care of the children. With rising affluence, parents can find many ways to lead fuller lives, but many of these opportunities must be forgone if the parents are burdened with the costs in time and money of raising children.

The agricultural and industrial revolutions, combined with the revolutions in sanitation and medicine, also transformed human health conditions, for children as well as for adults. Fewer died, and therefore fewer were necessary for a family to maintain itself. Perhaps equally important were improvements in communications, whereby new ideas and new attitudes could be rapidly disseminated, and the nature of family relationships changed. In the modern industrial society the family is no longer a little world unto itself, nor is it much protection against the larger world outside. Family ties are loose and tenuous in modern society. The old and new generations depend much less on each other. Children no longer represent much security for parents in their old age.

The causal relationships and the relative importance of these changes in bringing about the demographic transition in the developed countries are obscure. All we can say for certain is that there are powerful correlations between changes in birth rates and social and economic change. During the past two hundred years the presently developed countries have experienced a vast transformation in the human condition, which is called modernization, and it is clear that an integral part of this transformation was indeed the transition in human fertility behavior.

What seems important about most of the developing countries where birth rates are now falling is that they have all experienced marked social or economic improvement in the conditions of life of the poorer citizens. Korea, Taiwan, Singapore, and Hong Kong have had rates of economic growth that are among the highest in the world and the benefits of this growth have been widely disseminated throughout their societies. In Sri Lanka, economic conditions have been stagnant but social welfare has been greatly improved. Education is free to everyone in Sri Lanka, both to boys and girls, and literacy rates are high, over 80 percent, compared to less than 30 percent in the other countries of the Indian subcontinent. Health clinics, dispensaries, and hospitals are widely distributed throughout the country, and life expectancy for both men and women is over sixty-five years, in contrast to less than fifty years for India, Pakistan, and Bangladesh. Until very recently, every human being in Sri Lanka received a free ration of one kilogram of rice from the government once a week and a second kilogram at a nominal price.

Literacy rates are relatively high in all the poor countries where birth rates have come down so rapidly. Life expectancies are high and infant and child mortality relatively low. Communications have been greatly improved and there are real opportunities for social mobility. In most of these countries, per capita incomes have risen sharply, but more significantly, the distribution of income among the population is more equitable than in most of the countries where birth rates are still high. It is possible for the poor in these countries not only to meet their minimum physiological needs, but to have realizable aspirations for improvement in the conditions of life which will be easier to attain if they have fewer children. Moreover, effective public or private family planning programs have been underway for a decade or more. It is hard to tell how much of the decline in birth rates is due to the spread of modern methods of family planning and how much to the improvement in the conditions of life of the poor, but probably both factors are necessary for a substantial fall in birth rates.

Can the experience of this group of small, poor countries be a guide for the future elsewhere in the third world? In principle, the answer is clearly yes, although the obstacles to be overcome are very great, and the outcome may well depend on a level of effort in cooperation between the rich and the poor countries far greater than has hitherto been thought possible. But the alternative is so terrible to contemplate that the effort must be made.

SUGGESTED READINGS

Acsádi, G., and Nemeskéri, J. 1970. *History of human lifespan and mortality*. New York: International Publications Service.

Barnett, S. A. 1971. *The human species*. New York: Harper.

Freedman, Ronald. 1964. *Population, the vital revolution*. New York: Doubleday.

Hardin, Garrett. 1969. *Population, evolution and birth control*. San Francisco: Freeman.

Myrdal, Gunnar. 1968. *Asian drama: An inquiry into the poverty of nations*. New York: Pantheon Press.

National Academy of Sciences. 1971. *Rapid population growth: Consequences and policy implications*. Baltimore: The Johns Hopkins University Press.

Nortman, Dorothy. 1973. *Population and family planning programs, a factbook*, in Reports on population/family planning. New York: The Population Council.

Scientific American. 1974. *The human population*. San Francisco: Freeman.

20
SYNTHESIS
AND CHALLENGE

I hope this book has been a recitation of useful fact. But I reflect that, like any daily paper, it is also a catalog of troubles. In recent chapters particularly, we have seen man changing old troubles for new, with the inexorable march of culture from simple to complex and of societies from the country to the city.

The Man-Made Environment Revisited

Man has altered his relationship to the natural environment in two main respects: the food chain and the energy chain. In the course of these changes, by which he has profited enormously, man has altered his original ecosystem, sometimes beyond recognition. Settled agriculture has indeed disturbed the balance of nature, but it has greatly improved the health and increased the numbers of mankind. So, even more dramatically, have the two major technological developments of the past three hundred years: improving man's food supply and lowering death rates by various public-health measures.

Modernization and urbanization have proceeded hand in hand with improvement in health, as measured by rates of death and illness. The diseases of civilization have replaced those of primitive man; heart disease, strokes, cancer, peptic ulcer, and diabetes have replaced the infections and malnutrition of undeveloped countries. (Actually, we have seen

that primitive man enjoys better nutrition than overpopulated settled cultivators, as in Asia and Latin America.) For perspective, we might notice
that at the present time, sixty million people in the world die every year
(Figure 20.1). Half of these deaths, or thirty million, are of newborn
babies, infants, and toddlers in Asia, Africa, and Latin America. These un-
derprivileged children, whose resistance is undermined by malnutrition,
die from infectious diseases. About ten million, or one-sixth of the sixty
million annual deaths in the world, are from malaria, tuberculosis, and
other infections among adults. Another ten million are from cardiovascular
diseases and cancer; in other words, little more than one-sixth of all
deaths are due to the diseases of civilization. This is because less than
one-third of mankind lives in industrialized countries, where the major
determinants of health are income and age. Disease and death afflict
mainly the young in primitive and preindustrial societies and mainly the
old in modern, urban nations. Given the choice, we would certainly
prefer to die from heart disease at the age of seventy than from infection
and malnutrition in childhood. Table 10.3 gives mortality rates for infants under one year of age, showing a fiftyfold difference between developed and undeveloped countries. This is something to keep in mind during debates on the blessings versus the evils of science and technology.
And this is where the magnitude of differences should cause concern; we
need not beat our breasts because the United States is up or down a fraction of a percentage point in comparison with Denmark or Sweden.

When we look at developed countries like Britain and the United
States, we find that health is better in cities than in rural areas, whether

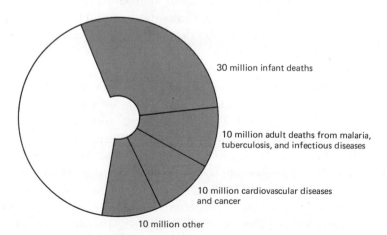

30 million infant deaths

10 million adult deaths from malaria,
tuberculosis, and infectious diseases

10 million cardiovascular diseases
and cancer

10 million other

20.1 Causes of yearly deaths around the world. (Combined figures from developed and underdeveloped nations.)

we compare rates of death or of illness. I have earlier presented some figures on physical illness. Even for mental illness, there seems to be little difference between city and country, in overall rates. The differences are not striking between country, small town, and city. The major differences in mental illness from one culture to another are not in frequency, but in particular manifestations. Within a single culture, particularly that of the developed, Western countries, the major differences in frequency are between social classes, or occupational levels. Mental illness increases steadily as one goes down the occupational scale, from professional to laborer. This is, of course, only an association. We do not know whether mental illness is a cause or an effect of social class. Do laborers have twice or three times as much mental illness as professional workers or executives because of their occupations and their socioeconomic status? Or, contrariwise, do families or persons with unstable personalities remain in or drift down to the lowest occupations? Finally, could the association reflect some third factor, common to both social class and mental illness?

These are important questions, but they are beyond our scope here. I do want to emphasize that modern life as such has not, as far as we can tell, increased the frequency of major mental illness in the population. Direct evidence of this point comes from Massachusetts. This state is one of the few places in the world where, for well over a century, there have been enough mental hospital beds to meet the needs of the population. Between 1840 and 1940, there was no marked increase in first admission rates for mental patients under age fifty. This is despite increasing industrialization and increasing urbanization in Massachusetts.

So we repeat: Modern life, and in particular city life, do not appear to have damaged human health. On the contrary, as cities and technology have spread, so have the life span and the well-being of the people living in such environments.

One might ask: could the positive or plus features of the city environment, such as sanitation, nutrition, public health, be outweighing its negative or minus features, such as noise, crowding, and poor housing? We have already seen that noise and crowding, by themselves, do not seem to do much real damage. Noise and crowding may be annoying, but they are not serious. The same seems to be true for poor housing. Many studies of the relationship between health and the urban environment, particularly housing, have appeared since 1920 and their results have been contradictory. Some investigators have found a direct, harmful effect of poor housing on health; others have reported no relationship; and still others have shown an actual increase in death rates or rates of neurosis, following rehousing of a slum population from an unhealthy area to much better housing!

Subtle Complaints about Cities. Therefore, if cities are such salubrious places, why do they continue to lose people who can afford to move to the suburbs? (I am not going to name names, but I can think of two prominent extollers—major authorities—of life in the city whose homes are most decidedly in rustic suburbs. And it would ill become me to throw stones; my own glass house is in Newton, another bedroom suburb.) The city, it seems, is a great place to visit or to work, but we wouldn't want to live there. And this is unquestionably a widespread sentiment. Why so? I suggest here that folk wisdom, or a kind of primordial instinct which is invoked by some authorities as well as laymen, or even simple individual perception and attitudes, may be more sensitive than our various present measures of health or illness. These, while pointing to better health in cities, may be too crude to detect the nonlethal and non-disabling, but nevertheless unpleasant and possibly harmful aspects of the urban environment. Distaste for the city may be based on aesthetic, philosophic, or psychological grounds, or on the very practical grounds of physical safety. But it cannot rationally be based on grounds of health. This notion is not purely speculative. There is evidence that, contrary to various fears, centuries of urban crowding need not be harmful. At least one sizable human group has made a successful biological adaptation to just such an environment—namely, the European Jews. Here is a group forbidden to own land for centuries, and crowded into unsanitary towns and cities under ghetto conditions. And yet, by all objective meaures of health and adaptation, they have done extremely well. They have low rates of death and illness and long life expectancy. They do not seem to have been harmed by their centuries of enforced estrangement from the beauties of nature. Now we certainly need more sensitive measures of environmental quality. In an unpublished paper, L. E. Hinkle has pointed out that the environment can have five kinds of effects on people: (1) It can shorten life. (2) It can impair bodily functions. These two effects we have already discussed. (3) The environment can impair mental or physical development, particularly among children. In this book, we have already seen how measures of child growth can detect very subtle effects of malnutrition or disease. (4) The environment can impose constraints on human behavior. (5) The environment can affect our feelings of pleasure and satisfaction. Although these five kinds of effect progress roughly from most serious to least serious, we cannot rank them in this way in respect to the importance they hold for people. People often persist in behavior harmful to their health, if they enjoy it. Examples would be overeating, alcohol consumption, cigarette smoking, and mountain climbing. So the five kinds of environmental effects must be evaluated separately.

We can evaluate environments by asking people what they think or by observing their behavior. Objective records, such as psychiatric clinic visits, school truancies and dropouts, or arrests for delinquency, would be examples of the kinds of behavior that could be correlated with environmental conditions.

It appears that any influence of the residential, or built environment on human health and behavior is not through direct physical or biological effects. The link is probably indirect and complex. It is probably mediated via social processes, perceptions, and attitudes, such as group cohesion, attitudes toward fellow-residents, tenant participation in decisions, ownership of dwellings, and the like. But we cannot be sure. This is a very important area for research. Few things could be more relevant to problems of the real world. We need to know much more than we do about the effects of the residential environment on health and behavior.

Man in the Global Ecosystem

The numbers, the technological efficiency, and the habits of consumption of modern man threaten our amenities of life and our supply of resources, whether these resources are renewable, like food and water, or nonrenewable, like minerals and fossil fuels. Our wastes and by-products impair the quality of our immediate environment and remotely threaten the entire ecosystem. Certain forms of animal life are in immediate danger. But for man, the danger is remote and potential, not immediate and actual.

No Free Lunch. Anything you do or refrain from doing has a cost (Figure 20.2). You do not get something for nothing. This principle of the trade-off is dramatically true of human ecology. Let us take a look at a few cost-benefit dilemmas.

We have already considered the technology versus nature dilemma. Everything man does affects his environment. It is impossible to obtain food and energy for large numbers of people without making a major impact on the natural environment. You may say, "We shouldn't have large numbers of people," but the argument comes too late. We have them; they are here already and will be for a long time. We must take as good care of them as we possibly can. This means large amounts of food, en-

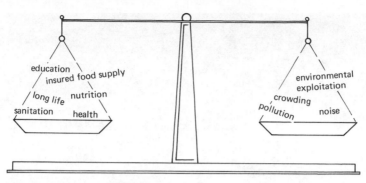

20.2   The trade-off. Everything has a cost.

ergy, and resources—even with present populations. To take one of many current estimates, the oil industry has said that world oil consumption will double in the next ten years. Almost as much oil will be used in the next decade as has been used in the last 110 years! To satisfy such a need, exploration would have to discover the equivalent of the Alaskan North Slope reserves every three months. Obviously, this situation cannot continue for very long; the most likely solution is alternative sources of energy, such as from sunlight or controlled nuclear fusion, unless we want to give up automobiles and electricity.

Take another dilemma—should we civilize primitive man? Anthropologists might have doubts. Some might like to preserve primitive cultures in a kind of human zoo, or game preserve, to be visited only by qualified scientists. But what about the people themselves (Figure 20.3)? Obviously, they welcome food, medicine, clothing, education, and the material benefits of civilization. Is it ethical to withhold medical treatment, when one dose of penicillin will cure the lifelong misery and eventual disfigurement of yaws? If you give medical treatment, can you ethically refuse further, continued treatment? When steel tools can reduce the labor of agriculture, should they be withheld? When Eskimos can avoid starvation and drowning by using firearms and outboard motors, should they be obliged to use harpoons and kayaks to satisfy our philosophic sensibilities or our admiration of their original cultural achievements, while we eat our three square meals each day and praise their simple, rugged life as seen on TV specials? The answer to all these questions, once again, is that we ask them too late. Primitive man has already encountered the larger culture and this contact will accelerate. Traditional ways of life are fast disappearing and in one or two more decades there will be virtually no functioning, intact primitive cultures left. This is enormously sad, in one sense. But from the point of view of the people them-

20.3 Solomon Island woman. Should fellow humans be denied the material benefits of civilization?

selves, biologically speaking, they will be much better off. Socially and culturally, they will not fare so well. The best we can hope to do, in the face of the inevitable, is to try to ensure that contact with our own culture is made by people who place the interest of the native people foremost. But who does? Missionaries? Traders? Settlers? Governments? We move on sadly to other dilemmas.

The Real Cost of Development. Possibly the most important general public health measure that this country could undertake would be to bring the pattern of living of the poor up to a level approximating that of the middle class. This will mean better sanitation, better nutrition, better housing, better maternal care, better child care, better medical care throughout life, and new patterns of adolescent activities and adult occupations. It will require millions of new housing units, thousands of new schools, possibly a 25 percent increase in electric power production, and some overall increase in the production of goods and services. All of this will create major demands upon the natural environment of the United States.

If even a minority of the underdeveloped countries of the world were to follow a pattern of rapid modernization, such as that pursued by Japan in the last three decades, and if this brought their populations up only to the level of health experienced by the relatively poor in the United States today, the demands upon the worldwide ecosystem would be very great indeed. It is clear, then, that any decisions that lead to restrictions in the use of natural resources or upon their distribution may have major implications for the health and welfare of many people. The impact of such restrictions will fall most heavily upon those who now have the least access to the fruits of modern civilization: namely, upon the poor and upon the people of underdeveloped nations.

Pesticides, Fertilizer, and Power: Dr. Jekyll or Mr. Hyde? Even the most circumscribed and apparently beneficial actions can carry with them important human implications. For example, there is now a widespread move against DDT, already banned in some areas. It has been pointed out that this persistent pesticide is responsible for the destruction of many birds and fish and that it might theoretically impair photosynthesis in the ocean. Although this is not borne out by studies of its distribution in the ocean, the reality of some of its bad effects seems to be clear.

There is also a widespread popular belief that DDT is an immediate threat to human health, but there have been few cases of clear-cut human disease even among people who are heavily exposed; and millions have been heavily exposed. Even though it is easy to find traces of DDT in human fat, the occurrence of human disease caused by chronic poisoning with the substance has not yet been established.

On the other hand, the banning of DDT would create a clear and immediate threat to human health. DDT has been responsible for the suppression of malaria in large areas of the world where this disease once caused 20 percent of all deaths and 10 percent of infant mortality. Its use to control malaria and other diseases, such as typhus, has saved hundreds of millions of lives in Southeast Asia, the Middle East, Africa, and Central America (Figure 20.4). Because of its cheapness and effectiveness there is

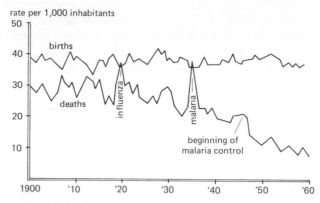

20.4   The birth and death rates in Sri Lanka. Against a constant birth rate, the death rate fluctuates with epidemics of malaria and influenza. The introduction of DDT to control malaria created a dramatic lowering of the death rate. (SOURCE: Population Reference Bureau. 1954. *Population Bulletin* 10:60–61; *American Journal of Public Health*, January 1972.)

no present substitute for it. If it were banned there would be an immediate cost in human illness and death. We do not have to speculate about this; it has been tried. In Ceylon, malaria had been endemic for thousands of years, causing incalculable misery. DDT, introduced in 1945, gave fifteen years of virtual freedom from this major killer. Its use was stopped in 1965. By 1968, three years later, there were over one million cases of human malaria in a population of ten million and no part of the island was free of the disease or its mosquito vector. In 1969, the Ceylonese government sent out an emergency call for ten million pounds of DDT to recover control. The same story would be repeated in many parts of the world if the use of DDT stopped. Suppose you were a public health official in one of those countries. How would you respond to a plea from the developed countries to stop DDT in order to preserve the Louisiana brown pelican or the peregrine falcon (Figure 20.5)? If DDT is indeed a serious threat to the worldwide ecosystem, some action should be taken against it but we must have the cause, the costs, and the facts clearly in hand as we decide upon the actions we will take.

It has also been said that the use of artificial fertilizers, containing large amounts of nitrogen, is a threat to human health. Some have implied that such fertilizers are causing widespread contamination of ground waters with nitrates and that this is a threat to health, producing an abnormal hemoglobin (methemoglobinemia) in large numbers of infants who are exposed to polluted drinking water. This is highly dubious.

On the other hand, it is quite clear what would happen if we were to

index of eggshell thickness

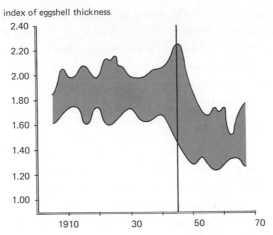

20.5 The peregrine falcon's egg showed a radical thinning of its shell following the introduction of DDT into the falcon's environment. The shaded area on the graph represents the range in thickness of the measured eggs, and the vertical line indicates when DDT was first introduced after World War II. (Redrawn from Ratcliffe, J. 1967. *Nature* 215: 208–210.)

stop using artificial fertilizers and other aids to agricultural production. Norman Borlaug, who won a Nobel Peace Prize for the Green Revolution in developing new strains of wheat, has pointed this out forcibly. Modern techniques of agriculture, including artificial fertilizers, herbicides, and pesticides, have doubled the per acre yield of food grains in the United States in the past twenty-five years. Each American farmer now feeds thirty-six other Americans and 6½ people overseas in countries like India and Egypt. It is easy to see what would happen to the health of many people if we stopped using these substances and our agricultural production fell abruptly. It is also easy to see to whom it would happen. It would happen to the poor in this country and to the desperately poor in countries overseas.

Once again, we do not need to guess; the experiment has been tried. In 1971, an agricultural specialist at the Virginia Polytechnic Institute compared yields from two adjoining garden plots, one grown organically and the other grown with the help of modern agricultural chemicals. To quote him, "The results were frightening. Insects and soil-borne disease organisms completely wiped out some of the vegetable crops, while causing low yields on others." To cite some of his examples, five rows of cucumbers yielded 205 pounds in the chemically treated plot, 29 pounds in the organic one; squash, 157 versus 3 pounds; eggplant, 155 versus 0 pounds.

Even a decision to build or not to build a power plant can have immediate human consequences. The power supply of many of our big cities is already marginal. Without even considering the consequences of major blackouts, like the one that occurred in New York City in 1965, one can state that even a transient power failure or cutting back of voltage could be very hazardous. A thousand people stalled in a subway train under the East River on a hot day is an invitation to a human catastrophe. A hospital which loses the power to its operating rooms and to the oxygen tents on the wards can produce human tragedies. During the British coal miners' strike in the winter of 1971–1972, millions of workers were laid off from factories which ran out of fuel and millions of homes were cold and dark.

It is now almost impossible to propose a new site to provide electric power anywhere without stirring up violent opposition. Hydroelectric plants dam rivers. This is opposed by fishermen and conservationists. Coal-burning plants cause air pollution. Mining coal, or anything else for that matter, destroys local landscapes. Drilling for oil, refining it, transporting it, and burning it cause problems. No community wants a nuclear-energy plant in its neighborhood. We are faced with some very hard choices. Do we want a nuclear power plant on the Hudson River (or elsewhere) in order to provide the people of New York City with electric power? Whom should we ask, the people who live along the Hudson or those in New York City? Should the decision be made by a head count? Nothing is free, least of all "power for the people."

The Equation of Death. It is sometimes stated that we might be more abstemious in our use of electric power. Air conditioning, for example, is a luxury we might do without in order to cut down on air pollution. But in any trade-off between the adverse human effects produced by air pollution from a power plant and the beneficial effects produced by using the power from that plant for widespread air conditioning, the power plant would win hands down. Summer heat waves far outweigh community-level air pollution as a cause of mortality in the United States (Figure 20.6). During the great weekend of smog around Thanksgiving 1966, there were no extra deaths in New York City. But during the heat wave that occurred in early July of the same year, 1,235 people died who statistically had not been expected to die. From a simple public health point of view, a contentious person might make a far better case for air conditioning all of the dwelling units in New York City than for cutting out all of the smoke from all of its power plants.

I was astonished to find how few people actually died in the famous air-pollution disasters so often quoted: 63 in the Meuse Valley of Belgium in 1930; 20 in Donora, Pennsylvania in 1948; 200 in New York City in

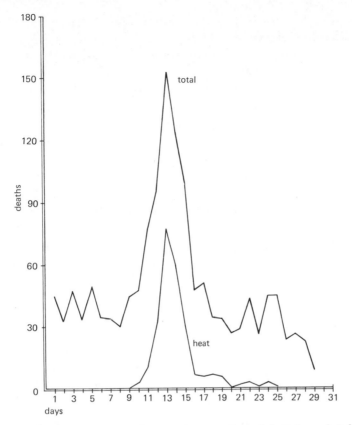

20.6 Daily deaths in St. Louis during July 1966. From July 9 through July 14, a heat wave occurred during which temperatures were an average 12–15° above normal (temperatures 90–95°). The upper line indicates the total number of deaths. The number of excessive deaths, those which would not have happened if there had been no heat wave, is shown by the smaller peak. (From Henschel et al. 1966. An analysis of the heat deaths in St. Louis during July 1966. *American Journal of Public Health*. 59:2232–2242.)

1953; 125 in 1967; and about 2,000 in three London smogs in 1952, 1956, and 1958. Most of these deaths occurred in elderly persons already suffering from heart or lung disease. (By way of contrast, 57,000 Americans lost their lives on the roads in 1970 alone.) Boston has never had an excess death from air pollution. To meet state laws, Radcliffe College switched from high-sulfur to low-sulfur fuel oil to reduce air pollution at a cost of $20,000 a year. That is four all-expense scholarships, each year. Harvard's shift could cost as much as $1 million a year. That is two hundred additional all-expense scholarships, each year, to reduce by a minute amount air pollution that has never been a threat to human health. Air pollution is objectionable for esthetic and economic reasons, but its role in human health is very small indeed.

Even such a mundane question as whether or not to build new storm sewers is not simply answered. At the present time, the storm drains of New York City are connected to the sanitary sewers. Even one-tenth of an inch of rain increases the volume of sewage so much that the treatment plants have to spill the excess sewage into the river. This is a highly undesirable practice from the point of view of the health of the river and the harbor. However, it does not create any significant amount of human illness. According to estimates, it will cost $2.4 billion over the next twenty years simply to modernize and repair the present sewer system and to build new sewers that will be needed. The cost of building an entirely new system of storm drains, separate from the sewers, would be much greater than this. New York City, as we well know, has been near the end of its financial tether. In some areas of the city, such as Harlem or Bedford-Stuyvesant, the infant mortality rate is double that in the suburbs such as Scarsdale and about 70 percent greater than that for the nation as a whole.

Two and a half billion dollars, or even a fraction of that, applied to the city's schools, prenatal care, and to infant nutrition might produce a significant increment in human health and welfare. A mayor with a limited health budget will have a hard time deciding that a much greater sum of money should be spent on storm sewers. Of course we must save the ecosystem. Of course we must start immediately to clean up our rivers, to protect our plants and animals, and to preserve our land. But in environmental affairs, as in all other human affairs, there are costs, which must be weighed carefully at each step. Some of them are very important human costs and these cannot be swept under the rug.

Here is another example of cost-benefit dilemmas. In 1955, the need for an additional airport for the New York City area became apparent, from increasing delays and near-accidents at the three existing airports, LaGuardia, Kennedy, and Newark. A remote site was found, near Morristown, New Jersey, with very few residents, mostly wealthy persons living on large estates. These residents objected, as everyone does, to living near an airport. So do the people of East Boston and Cape Cod. By having the Great Jersey Swamp declared a National Wildlife Refuge, they blocked the airport. Everybody likes a wildlife refuge. But as a result, seven hundred thousand people living in Jamaica, Long Island, near the existing airports, continue to have their sleep disturbed nightly and one hundred thousand school children have their classes disrupted many times during the day by airplane noise. Long delays occur around airports and near-accidents increase in number. How would you settle this problem? By reducing the amount of air traffic? How many of us are planning to walk or cycle home for our Christmas vacations or to row or swim to Europe next summer?

Let me give a further example. The Everglades, in Florida, are a priceless natural resource. For many years, the water level in the Everglades has been lowered by drainage to provide land for orange and grapefruit groves. I deplore this. At the same time, consider the contribution that citrus fruit and its products, in large quantities, at prices within the reach of all have made to the health of children in the United States and around the world. Everything has a cost.

In trying to solve some of those problems, we clearly need a method of analysis which will allow us to consider ecosystems as just that—systems. Whatever you do has effects on other components in the system. There is indeed a technique for analyzing such interactions, called systems analysis, a branch of communication theory, that is being used increasingly in engineering and economics. The systems approach might have prevented the fiasco of the Aswan High Dam in Egypt. Although built at enormous cost and with the best of intentions, it may end up doing more harm than good. It turns out that more than half of the Nile water impounded behind the dam is lost by evaporation. Another sizable fraction is lost by seepage into the surrounding porous limestone. Millions of tons of silt, rich in nutrients, sink to the bottom of the dam and are no longer poured into the Eastern Mediterranean. This has destroyed the fisheries of that area, formerly a major source of food. Instead of an alternating cycle of dry and wet seasons, with the snail carriers of schistosomiasis dying in large numbers during the dry season, there is now constant irrigation. This spreads the snails, which carry the worms, over greater and greater areas. It is estimated that between 55 and 70 percent of the Egyptian population will become infested, making a total of at least seventeen million cases. While the Aswan dam will increase by one third the land available for cultivation, its medical costs alone may offset the agricultural gains. And the irony is that during the time of construction and operation of the dam, the Egyptian population, which doubles in twenty-four years, will have expanded to the point of just balancing the gain in arable land—quite apart from medical costs and the loss of fish. The net result will be to have a large number of people on the verge of starvation instead of a smaller number.

How Do We Decide? The other main need in solving our ecological problems is for a better understanding of human motivation. Everyone is in favor of personal freedom and against coercion. But how much freedom

can be allowed in matters of health and survival? Who would be vac-
cinated against smallpox or take a blood test for syphilis if he didn't have
to? Who would fluoridate his water supply? Not many towns have voted
for it. In fact, not many towns and cities vote for education either, these
days. Who will give up his land for waste disposal facilities or pay taxes for
expensive, smokeless incinerators? Who will pay the increased cost of
low-sulfur fuel, once the cost becomes apparent?

Participatory democracy, in which everyone affected by a decision
must be consulted and must concur in the decision, can paralyze action
needed for the common good. The classic example is the city of Calcutta,
in India, long before the recent Bangladesh refugee problem. For years,
the greatest need of the city has been for sewer construction. But nobody
will give up his land. Would you? As a result, no sewers have been built
for many decades, the population keeps increasing, and even without the
added burden of refugees, the city lives on the brink of an epidemic of
terrifying proportions.

Some of the gloomier writers on population, like Kingsley Davis,
Garrett Hardin, and Paul Ehrlich, are approaching the position that reproduction may have to be controlled in the greater interest of all. I
believe this is unnecessary, as shown by the continuing fall in birth rates
in the advanced countries. Several, including the United States, are dip-
ping below replacement levels. But I must confess that my optimism can
be criticized as unrealistic on a global scale, considering the time constraints under which we are operating.

In any case, we need more sophisticated analyses of motivation than
we now have. It turns out that by a rational calculation people are justified
in taking risks when these risks are in the future and are small in degree,
whereas avoiding them costs you something here and now. For example,
the the chance that a heavy smoker will die of lung cancer is one in a
thousand; the chance for a nonsmoker is one in ten thousand. One in a
thousand is a pretty small risk, far in the future. If smoking gives you
pleasure, you may accept the risk. Many programs to promote public
health based on scare education have failed because of such calculations.
We need to devise methods of motivating people which they will perceive
to be in their own immediate, short-term interest. This may have to take
the form of bribery to conform, or penalty for not conforming. A little
sugar helps the medicine go down.

Finally, what of the future? Can we protect the environment and
provide for people at the same time? We have no choice but to try. I am
fundamentally an optimist. I believe that we can do this, by limiting populations; by acquiring abundant sources of energy, free of pollution, such

as from sunlight or from new techniques of nuclear fission or fusion; and by substituting synthetic for natural products in many manufactured articles. We shall have to recycle, to dispose of human and industrial wastes wisely, and to consider the entire globe as one ecosystem. Pollution is a worldwide aspect of technology, not a capitalist perversion. Russian rivers burst into flame, just as ours do. Lake Baikal, in Russian Siberia, one of the natural wonders of the world, is fast becoming a dead Lake Erie. Public greed is no more considerate of the environment than private greed. But human intelligence and adaptability should be equal to the task, challenging though it is.

And now, in conclusion, let me summarize briefly what this book has tried to do. The first third was concerned with human evolution and genetics—the stages and the mechanisms whereby man reached his present biological status as a species. In the middle third, we considered mankind today—specifically, human variation as expressed in age, sex, race, and individual constitution. I described man's physical, biochemical, and physiological characteristics, his adaptation to various environments, and his suceptibility to various diseases, and I outlined the ongoing trends in his evolution. Finally, in the last third, I have been concerned with human ecology—how the environment affects man and how he in turn adapts to and alters all his environments—primitive, urban, and global.

Beyond the facts, I hope I have indicated, in one way or another, some of the broadest aspects of human biology: what is important to know; what little we do know; and how we might go about finding out what we need to know.

Challenge

I believe, and hope readers will agree, that man's relation to his ecosystem is an area of immense importance. The biological effects of various environments on man should be a major input to any long-range planning, on any scale—local, urban, regional, national, and international. A main challenge is to order our priorities. Save your indignation, your effort, and your emotional investment for the big issues—population and resources, hunger and poverty. We should not fritter them away on air pollution or heart transplants.

I hope that in thinking about human biology and ecology, you, the reader, will prize evidence in addition to emotion and that you will see the need for competence as well as concern. Look at issues in an objective, factual, and quantitative way. Ask yourself, "How many people are affected and how severely? How do we know? How do we find out?" These are the questions that guide a research scientist in the biology of man. But they are just as important for the educated layman, who will be thinking and reading about the human future all his life.

GLOSSARY

Adaptation: any characteristic that aids an organism to survive and reproduce in the environment it inhabits. If the adaptation is genetically controlled then it is important to evolution.

Adaptive radiation: the evolutionary branching-out from a basic animal form in following divergent lines of adaption.

Adenine: a purine base found in DNA and RNA.

Allele: any of several gene forms that can occur as alternates at a specific locus on a chromosome.

Allen's rule: other things being equal, warm-blooded animals of the same general kind will have shorter extremities in the colder parts of their range than animals in the warmer parts.

Amino acids: complex organic compounds carrying an animo group $(-NH_2)$; the building-block compound of proteins.

Androgen: general term for male sex hormones.

Assortative mating: as opposed to random mating; like tends to pair with like (positive) or unlike with unlike (negative).

Autosome: any chromosome other than a sex chromosome.

Balanced polymorphism: a condition of equilibrium among different alleles at a given chromosomal locus. In a population having individuals of two or more genetically distinct types, the relative frequencies of the types remains constant from generation to generation.

Barr bodies: sex-chromatin bodies; deeply staining particles found attached to the inner surfaces of the nuclear membranes of the cells of females. According to the Lyon hypothesis, this body is formed by one of the X chromosomes which is inactive.

Bergmann's rule: other things being equal, warm-blooded animals of the same general kind will be larger in general body size in the colder parts of their range.

Blood group: the classification of types of blood based initially on the occurrence of the clumping of the red blood cells when blood of incompatible groups is mixed.

Centriole: a cytoplasmic body, located just outside the nucleus, that organizes the spindle during mitosis and meiosis.

Centromere: the region of the chromosome that attaches to a spindle during mitosis and meiosis.

Chromosomes: thread-like structures in the nuclei of cells, on which genes are located. They are paired (one coming from each parent) and, for each species, are constant in number. Man has twenty-three pairs.

Cline: gradual geographic variation in a characteristic of a species.

Codominance: the independent expression of two alleles in the heterozygote, an example of which is the *AB* blood group of the *ABO* system.

Congenital: existing at birth but not necessarily hereditary.

Crossing-over: exchange of parts of two homologous chromosomes.

Cytosine: pyrimidine base found in DNA and RNA.

Deletion: a form of mutation in which a portion is lost (deleted) from the chromosome.

Diploid: the chromosome state in which each type of chromosome is represented twice, or in pairs ($2n$), in contrast to haploid ($1n$).

Dizygotic twins: two zygotes; twins which develop from separate ova; fraternal twins (see monozygotic twins).

DNA: deoxyribonucleic acid; a nucleic acid found in the cell nucleus, thought to be the genetic material.

Dominant: of an allele or by extension a trait. An allele is dominant if its phenotypic effect is expressed in both the homozygous and heterozygous condition.

Drift: see genetic drift.

Ecosystem: the sum total of physical features and organisms occurring in a given area and their interactions.

Ectomorph: one of three idealized body forms, lean and slightly muscular. See endomorph and mesomorph.

Endomorph: one of three idealized body forms, tending to fat. See ectomorph and mesomorph.

Enzyme: a protein which functions as a catalyst. It is made by cells to accelerate chemical reactions.

Epidemiology: the study of the factors involved in the distribution and frequency in infectious diseases.

Epiphysis: a joint surface or process of a bone which ossifies and grows separately, and unites with the body of the bone near the time of maturity; this is characteristic of mammals.

Equilibrium, genetic: a condition in which a population remains the same genotypically through time.

Estrogen: one of a group of vertebrate female sex hormones.

Fetus: unborn young; an embryo in its later development, still in the uterus.

Founder effect: when emigrants from a population do not possess a random sample of the genes of that population, the colony will differ genetically from the ancestral population.

Gamete: a mature sex cell (egg or sperm) with a haploid chromosome set.

Gene: the unit of inheritance. A portion of DNA within a chromosome.

Gene flow: the spread of genes from one breeding population to another or from one part of a population to another, by interbreeding.

Gene pool: the sum total of all the genes of all the individuals in a population.

Genetic drift: fluctuation of gene frequencies through chance rather than selection, mutation or migration.

Genotype: the particular combination of genes in the chromosomes of an individual. See phenotype.

Germ cell: a sexual reproductive cell; an egg or sperm.

Guanine: a purine base found in DNA and RNA.

Haploid: the chromosome state in which each type of chromosome is represented only once ($1n$) in contrast to diploid ($2n$)—the number of chromosomes found in a germ (egg or sperm) cell, following meiosis.

Hardy-Weinberg law: in large random mating populations, gene frequencies will remain the same over time in the absence of mutation, migration or selection.

Hemoglobin: the oxygen-bearing iron containing protein found in red blood cells.

Heterosis: a condition in which hybrids between distinct strains are thought to be superior to both parent strains in some respects; hybrid vigor.

Heterozygous: having two different alleles of a given gene group. See homozygous.

Homo erectus: species of early man living before about three hundred thousand years ago, remains of which have been found in Europe, Africa, and the Orient.

Homozygous: having two identical alleles of a given gene. See heterozygous.

Hormone: a chemical messenger or central substance secreted into the bloodstream by the endocrine glands. Hormones exert control over aspects of bodily development and metabolism.

Immunoglobulin: a group of related serum proteins to which antibodies (such as gamma globulin) belong.

Inbreeding: the continued breeding of successive generations of close relatives.

Independent assortment: during meiosis, in the formation of gametes, the distribution of the members of one pair of genes does not influence the distribution of the members of other pairs.

Insertion: transposition of chromosome fragment to a nonterminal location within a chromosome.

Karyotype: the chromosome complement of a somatic cell representing an individual or species. The term is often applied to diagrams of chromosomes as they are lined up in homologous pairs.

Keloid: hyperplastic fibrous connective tissue, usually at the site of a scar.

Keratin: a fibrous sulfur-containing protein forming the outer layers of such epidermal structures as hair, nails, and horns.

Klinefelter's syndrome: in males, a genetic abnormality caused by the possession of two X chromosomes and one Y chromosome (XXY); characteristics are of a male but the testes are not functional. The individual may have feminine breast development.

Longitudinal study: a study which follows a trait during the life of an individual.

Lyon hypothesis: in any given cell in a female, one X chromosome is active, the other is inactive and forms the Barr body.

Meiosis: the process by which the chromosome number is halved to produce gametes (sperm or eggs) having a haploid chromosome set.

Melanin: dark pigment found in cells of the iris of the eye, as well as skin and hair of vertebrates including man.

Melanocyte: an epidermal cell capable of synthesizing melanin.

Menarche: the first occurrence of menstruation in girls.

Mesomorph: one of three body forms characterized by powerful muscle and a predominantly bony frame. See ectomorph and endomorph.

Metabolism: the total of the chemical processes carried on in the body including digestion, respiration, secretion, excretion release, and utilization of energy. The chemical processes in living cells by which energy is provided, new materials assimilated or synthesized and wastes removed.

Millirem: one-thousandth of a rem.

Mitosis: the process by which a cell is divided into two daughter cells each having a complement of chromosomes identical to that in the parent cell.

Mongolism (Down's syndrome): deficiency is accompanied by, among other things, short stature, stubby fingers, and a fold on the eyelid.

Monozygotic: one zygote, referring to twins or other multiple births which develop from a single ovum—identical twins, triplets, etc.

Morbidity: sickness trend in a population. As used here morbidity refers to disease rates and types.

Mortality: death trend in a population. The mortality rate refers to the number of people who die, usually per thousand.

Mutation: a sudden heritable alteration of genetic material involving quantity, quality or arrangement. (Generally, mutation refers to a permanent change to the genotype; in descendants which receive the mutant gene there is a changed effect to which the mutant gene is related.)

Natural selection: differential reproduction leading to an increase in the frequency of some genes and the decrease of others. It is caused by certain individuals who are more adapted for a given situation passing on the genes responsible for their favorable adaptation to the next generation in a greater proportion than the remainder.

Nondisjunction: the failure of a homologous pair of chromosomes to segregate at metaphase so that one daughter nucleus receives both chromosomes of the pair and the other daughter nucleus receives neither. Nondisjunction can occur during meiosis and mitosis.

Nucleic acid: a class of chemical compounds formed by the combination of purine and pyrimidine bases with a sugar and phosphoric acid.

Nucleotide: the basic unit of nucleic acid consisting of sugar molecule (deoxyribose or ribose), a phosphate, and a purine and pyrimidine base (cytosine, thymine, guanine, adenine, or uracil).

Peptide: a compound containing two or more amino acids. Peptides join to form polypeptides which in turn make proteins.

Phenotype: the physical expression of a genetic trait. The appearance of an individual.

Pleiotropic gene: any gene which has more than one phenotypic effect.

Pleiotropy: the multiple phenotypic effects of a single gene.

Point mutation: the localized alteration of the base sequence of DNA molecule comprising a gene.

Polygene: a gene which by itself has small effect on the phenotype but in conjunction with several or many genes control a quantitative character (height, weight, degree of pigmentation, etc.).

Polymorphism: "many forms": the occurrence of different morphs (e.g., blood types, colors, hair forms) in members of the same species. Frequently the term is applied to genetic variation among individuals within one race.

Polypeptide chain: peptides linked together. See peptides.

Primary sex ratio: the male-female ratio of zygotes. The ratio at the time of conception rather than at birth. See sex ratio.

Primates: an order of mammals to which man, apes, monkeys, and lower primates belong. (The lower primates include lemurs, lorises, and tree shrews.)

Proteins: compounds composed of amino acids forming the most important structural and enzymatic constituents of the body. A long polypeptide chain.

Pulmonary: referring to the lungs.

Pyrimidine: any of a class of organic (nitrogenous) bases found in nucleic acids.

Radiography: producing an image on a radio-sensitive surface by radiation other than of visible light; especially X ray.

Recessive: referring to a gene that does not produce a phenotypic effect in a heterozygote but can only have expression as a homozygote. A trait that is shown phenotypically only by homozygotes.

Rem: from Roentgen Equivalent Man. A unit employed in measuring radiation. Specifically, the amount of any ionizing radiation that will cause the same amount of biological damage to human tissue as one roentgen.

Ribosome: small particles in the cytoplasm of a cell that are the sites of protein synthesis. Messenger-RNA attaches to them.

RNA: any of several nucleic acids in which the sugar component is ribose and one of the nitrogenous bases is uracil.

Roentgen: unit commonly employed in measuring the amount of radiation, based on the amount of ionization produced.

Secondary sex ratio: the male-female ratio at birth.

Sex chromosomes: chromosomes that exert genetic control over sex determination. *XX* in women and *XY* in men.

Sex linkage: having the gene carried in the sex chromosomes (usually the *X*).

Sex ratio: the relative proportion of males and females in a population. By custom, the ratio presents males as a percentage of females; that is, the number of males per hundred females.

Somatic cell: refers to body cells and tissues as opposed to germ cells.

Somatotype: the morphological type of a human body. See ectomorph, endomorph, and mesomorph.

Stabilizing selection: selection which in a large population discriminates against mutations that produce phenotypes outside a range conducive to the success of the species.

Steatopygia: exceptional fat deposits on the buttocks. This condition is found especially among women of the Hottentots and Bushmen.

Thymine: one of the organic (pyrimidine) bases found in DNA. The analogous base in RNA is uracil.

Trisomy: having an extra chromosome added to any pair, one kind of chromosome is present in triplicate. When the twenty-first chromosome is affected in man (trisomy 21) the result is Down's syndrome (mongolism).

Turner's syndrome: Genetic abnormality in human females caused by the presence of only one sex chromosome (XO). The characteristics are those of a female usually of short stature, often retarded, and always sterile.

Uracil: a pyrimidine base found only in RNA. The analogous base in DNA is thymine.

Vertebrates: the major group of animals which possess spinal columns, from fish to mammals. Technically a subphylum of the phylum Chordata.

X chromosome: a sex chromosome of which women normally possess two and men one.

Y chromosome: a sex chromosome of which men normally possess one and women none.

Zygote: the chromosome complement of a newly fertilized egg. The result of the union of two gametes, the egg and the sperm.

INDEX